Uganda

NATIONS OF THE MODERN WORLD: AFRICA
Larry W. Bowman, *Series Editor*

UGANDA
Tarnished Pearl of Africa

THOMAS P. OFCANSKY

WestviewPress
A Division of HarperCollinsPublishers

Nations of the Modern World: Africa

Published in 1996 in the United States of America by Westview Press, Inc., 5500 Central Avenue,
Boulder, Colorado 80301-2877, and in the United Kingdom by Westview Press, 12 Hid's Copse Road,
Cumnor Hill, Oxford OX2 9JJ

Library of Congress Cataloging-in-Publication Data
Ofcansky, Thomas P., 1947–
    Uganda : tarnished pearl of Africa / Thomas P. Ofcansky.
      p.   cm. — (Nations of the modern world. Africa)
    Includes bibliographical references and index.
    ISBN 0-8133-1059-8 (cloth) — ISBN 0-8133-3724-0 (pbk)
    1. Uganda.   I. Title.   II. Series.
DT433. 222. O35   1996
916. 761—dc20   95-13468
                                                                                  CIP

The paper used in this publication meets the requirements of the American National Standard for
Permanence of Paper for Printed Library Materials Z39. 48-1984.

10   9   8   7   6   5   4   3   2   1

"But the Kingdom of Uganda is a fairy-tale. You climb up a railway instead of a beanstalk, and at the end there is a wonderful new world."
—*Winston S. Churchill*
My African Journey, *1908*

Contents

4 SOCIETY AND CULTURE   69

5 THE ECONOMY   93

6 INDEPENDENCE AND
NEW FOREIGN POLICY DIRECTIONS   125

7 THE UGANDAN EXPERIENCE AFTER
THREE DECADES   153

Contents   ix

Illustrations

Preface

My interest in East Africa stretches back over a period of more than twenty years. During that time, I have traveled throughout the region and have studied topics as diverse as wildlife conservation; the British colonial period; military history; human rights; the social, political, and economic impact of famine and large-scale refugee migrations; and the proliferation of arms and guerrilla groups. Throughout these endeavors, Uganda has always held a special fascination for me, largely because of my inability to reconcile the disparities in that country. Uganda is lush and beautiful, and its people are kind and gentle. However, much of the country's postindependence history has been characterized by unimaginable violence and brutality. Although writing this book has not stilled my curiosity about the contradictions in Ugandan society, it has afforded me the opportunity to record some of my own impressions of the country and to assess what the future holds in store for all Ugandans.

Many people have supported my efforts during the preparation of this volume. Robert O. Collins, a longtime friend and colleague from the Department of History at the University of California–Santa Barbara, read an early draft of the manuscript and offered many helpful comments. Roger Winter, director of the U.S. Committee for Refugees, also read the manuscript and allowed me to use many of his photographs of Uganda. Additionally, Jacqueline DeCarlo, project coordinator of the U.S. Committee for Refugees, assisted with the search for photographs. Rodger Yeager from the Department of Political Science at West Virginia University kindly reviewed the manuscript prior to its publication. During a recent trip to Uganda, Ellen Shippy, Thomas and Jennifer Underwood, and Thomas Spang, all of whom worked at the U.S. Embassy in Kampala, were generous with their time and hospitality. They also facilitated my understanding of the many problems confronting Yoweri Museveni's government. I would also like to thank Larry W. Bowman, general editor of the Nations of the Modern World: Africa Series, for his many thoughtful comments and criticisms. Barbara Ellington, acquisitions editor at Westview Press, ensured that my efforts always remained focused on this project. Jane Raese, production editor at Westview, used her considerable skills to prepare the manuscript for publication. To all of these individuals I owe a debt of gratitude.

Thomas P. Ofcansky

Acronyms

ACP	AIDS Control Program
AIDS	Acquired Immune Deficiency Syndrome
BADEA	Banque Arabe de Développement Économique en Afrique
BAT	British American Tobacco
CCIA	Central Council of Indian Associations
CMS	Church Missionary Society
CNN	Cable News Network
COMESA	Common Market for Eastern and Southern Africa
CP	Conservative Party
DP	Democratic Party
EAC	East African Community
EC	European Community
EPRC	Education Policy Review Commission
ERP	Economic Recovery Program
FAZ	Forces Armées Zairoises
FEDEMU	Federal Democratic Movement of Uganda
FIDA-U	Uganda Association of Women Lawyers
FO	Foreign Office
FUNA	Former Uganda National Army
GDP	Gross Domestic Product
HIV	Human Immunodeficiency Virus
HSM	Holy Spirit Movement
IBEAC	Imperial British East Africa Company
ICO	International Coffee Organization
IGADD	Inter-Governmental Authority on Drought and Development
IGG	Inspector general of government
IMF	International Monetary Fund
IUCN	International Union of Conservation of Nature and Natural Resources
KAR	King's African Rifles
KBO	Kagera Basin Organization
KY	Kabaka Yekka
LRA	Lord's Resistance Army
MNC	Mouvement National Congolais

NALU	National Army for the Liberation of Uganda
NAM	Non-Aligned Movement
NCC	National Consultative Council
NCHE	National Council for Higher Education
NGO	Nongovernmental organization
NRA	National Resistance Army
NRC	National Resistance Council
NRM	National Resistance Movement
OAU	Organization of African Unity
PFLP	Popular Front for the Liberation of Palestine
PJMCC	Permanent Joint Ministerial Commission of Cooperation
PLC	Parti de Libération Congolaise
PLO	Palestine Liberation Organization
PP	Progressive Party
PRC	People's Republic of China
PTA	Preferential Trade Area for Eastern and Southern Africa
RC	Resistance Council
RDP	Rehabilitation and Development Plan
RPF	Rwanda Patriotic Front
SAF	Sudanese Air Force
SPLA	Sudanese People's Liberation Army
TPDF	Tanzania People's Defence Force
UAC	Uganda Airlines Corporation
UAFU	Uganda African Farmers' Union
UAG	Uganda Action Group
UCP	United Congress Party
UDC	Uganda Development Corporation
UDCM	United Democratic Christian Movement
UFM	Uganda Freedom Movement
UHRA	Uganda Human Rights Activists
UIA	Uganda Investment Authority
UN	United Nations
UNAMIR	United Nations Assistance Mission in Rwanda
UNC	Uganda National Congress
UNDP	United Nations Development Program
UNFPA	United Nations Fund for Population Activities
UNHCR	United Nations High Commissioner for Refugees
UNICEF	United Nations Children's Fund
UNLA	Uganda National Liberation Army
UNLF	Uganda National Liberation Front
UNM	Uganda National Movement

UNP   Uganda National Parks
UNRF   Uganda National Rescue Front
UP&TC   Uganda Posts and Telecommunications
UPA   Uganda People's Army
UPC   Uganda People's Congress
UPDA   Uganda People's Democratic Army
UPDM   Uganda People's Democratic Movement
UPM   Uganda Patriotic Movement
UPU   Uganda People's Union
URC   Uganda Railways Corporation
USAID   United States Agency for International Development
UTC   Uganda Transport Company
UWLA   Uganda Women's Lawyers Association
WHO   World Health Organization

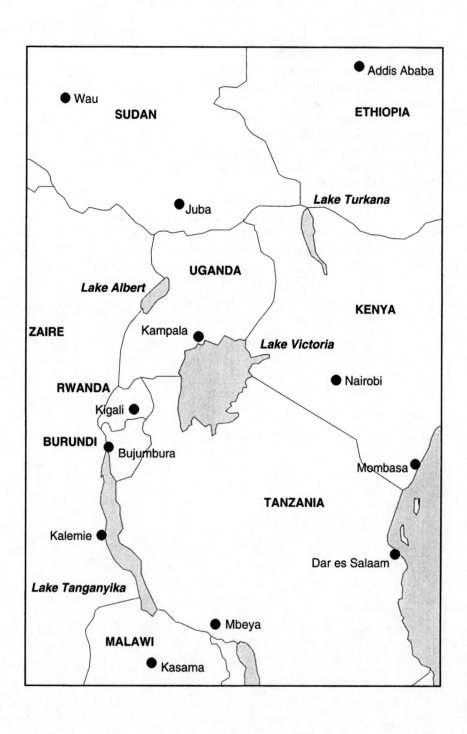

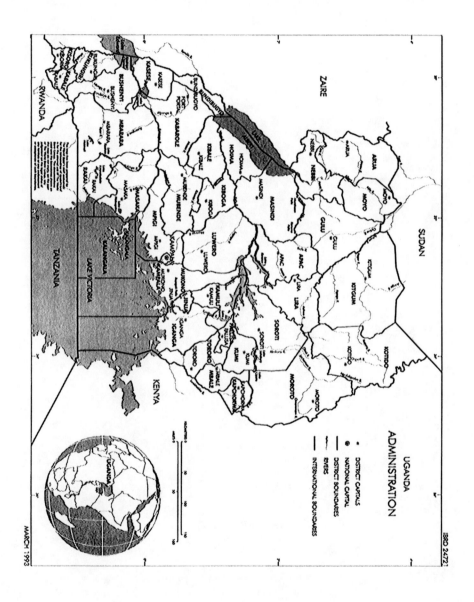

UGANDA
ADMINISTRATION

DISTRICT CAPITALS
NATIONAL CAPITAL
DISTRICT BOUNDARIES
RIVERS
INTERNATIONAL BOUNDARIES

KILOMETERS 0   50   100   150
MILES 0   50   100

MARCH 1993

IBRD 24721

INTRODUCTION

THIS BOOK PROVIDES an overview of Uganda, a country that represents the hope and despair of modern Africa. The study begins with a brief examination of the factors and themes that have influenced Uganda's historical development. However, the book will focus mainly on the postindependence period. During this era, Uganda gained the reputation of being one of Africa's most violent and politically troubled countries. After seizing power in 1986, Yoweri Museveni promised to restore stability, rebuild the economy, and institute political reforms and a democratic form of government. Efforts to achieve these goals have provoked widespread controversy among scholars, government officials, and humanitarian workers.

Those taking a sympathetic point of view argue that the Museveni regime represents a fundamental change in the character of Uganda's political leadership. According to this interpretation, Museveni and his National Resistance Movement/Army (NRM/A) have provided Uganda with its best and most effective government since independence. Pro-Museveni supporters defend this judgment by pointing to the fact that much of the country is at peace, there is respect for human rights and freedom of speech and the press, there is a functioning local government, and the economy is on the mend. Because of these and other achievements, many sympathetic observers maintain that the current Ugandan government could be viewed as a model for other African states struggling to improve the welfare of their peoples.

Critics, however, admonish Museveni for refusing to sanction multiparty elections, failing to stop government corruption, and pursuing an aggressive foreign policy, which many neighboring nations interpret as little more than an attempt to impose a Pax Uganda on eastern Africa. Advocates of this view contend that despite some recent political and economic improvements, Uganda remains a deeply divided society that could easily unravel after the end of the Museveni presidency.

Despite the controversy over the direction of Museveni's revolution, there is general agreement that Uganda is undergoing a unique experiment that will determine the nature and scope of its political, economic, and social life well into the next century. Whatever the future holds in store, Uganda will remain a symbol of all that is evil and all that is promising in the human spirit. In this respect, it typifies many African countries that are trying to come to terms with divisive elements that presumably could be harnessed to develop better societies. But all too often these elements have degenerated into the worst sort of depravity, which in turn caused suffering and hardship on unprecedented scales.

To appreciate the challenges confronting contemporary Uganda, the reader must understand some basic facts about the country. Despite its agricultural potential, Uganda is desperately poor and the economy is overdependent on declining coffee revenues. Although it has been improved in recent years, the country's transportation infrastructure is substandard and is one of the country's major economic bottlenecks. Similarly, government services at all levels are minimal; in some remote areas of Uganda they are nonexistent. The presence of large refugee populations, which at any given time can number in the tens or hundreds of thousands, further strain the country's limited resources. Although precise figures are unavailable, it is known that unemployment and underemployment rates—especially in cities, towns, and villages—are high among all age groups. As a result, many Ugandans with professional skills go abroad if possible. Other Ugandans who lack education or training have turned to banditry or other forms of crime merely to survive. Looming over everything in Uganda is the Acquired Immune Deficiency Syndrome (AIDS) epidemic, which threatens the well-being of the entire population.

Neither Museveni's regime nor any future Ugandan government will possess the resources necessary to resolve these problems and to break the destructive cycles of poverty, disease, crime, instability, violence, and refugee migrations. If anything, the country's human population—which continues to grow despite the spread of AIDS—will slowly exacerbate the difficulties confronting Ugandan society. By the next century, the gulf between a Ugandan's expectations and reality will be wider than ever.

This prognosis should come as no surprise to those who follow African affairs, as these problems plague almost every nation on the continent. For the generalist, however, this outlook may seem excessively pessimistic. No matter how it is interpreted, this book seeks to provide the reader with the facts necessary to understand one of Africa's most fascinating nations.

1

THE PHYSICAL SETTING

U GANDA IS A LANDLOCKED COUNTRY located along the equator, about 500 miles from the Indian Ocean. The country is bordered by Sudan in the north, Kenya in the east, Tanzania and Rwanda in the south, and Zaire in the west. Uganda's total area is 91,076 square miles, including 16,364 square miles of open water or swamp. The country is largely situated on a plateau 3,000 to 5,000 feet above sea level, dissected by many rivers, swamps, and lakes. Mountain ranges and other relief features are located on the borders. Uganda's western region contains the western Rift Valley, the Mufumbiro Volcanoes, and the Ruwenzori Mountains, which are also known as the Mountains of the Moon. Lake Victoria constitutes much of the country's southern boundary, and the highlands, which are dominated by volcanic Mount Elgon, mark most of the eastern boundary.

Mineral Resources

Uganda possesses a variety of commercially valuable minerals.[1] Copper, first discovered in 1906, is found at Kilembe, which is located in a valley among the Ruwenzori Mountains' lower eastern slopes. There is a tin belt that reaches a width of 40 miles and runs northwest to southeast between two gold belts. The center line of the tin belt is located where the Ruizi River turns southward to join the Berarara River. Mwirasandu was the country's largest mine, but operations stopped after thirty years because the deposits had been exhausted. Low-level tin mining is still possible in other parts of the southwest. Beryllium ore, used for nuclear energy and as an alloy with copper and other metals, is also found in the southwest. Although deposits are sufficient to allow Uganda to become a beryllium-producing country, a large-scale export market has never been developed.

Large gold deposits are located in two belts running northwest to southeast. The western gold belt, which has a maximum width of about 20 miles, passes near

Kabale. The center line of the eastern belt, which is approximately the same width, is west of Mbarara. Lesser amounts of gold can be found in the north near Kitgum and around Moyo. There are four low-grade tungsten (wolfram) mines in Kigezi District. However, historically unstable world prices have prevented Uganda from realizing a consistent profit from these mines.

Other minerals present in Uganda are columbium and tantalumores (Tororo); bismuth (Kigezi District); lithium (west of Lake Victoria); silver (Kitaka); galena (Kitaka); mica (Labwor Hills, West Nile District, and south of Kampala); and chromium (Karamoja). Additionally, the country possesses a variety of high-quality iron ores, including hematite ores (Kigezi and northeast Ankole); magnetite (Sukulu and Bukusu); thorium (Karamoja); barites (Ankole and Karamoja); asbestos (West Nile, Acholi, and Karamoja); diatomite (west of Katonga River); topaz (East Mengo); and graphite (Karamoja and West Nile). There are also garnet, talc, feldspar, gypsum, vermiculite, and potash deposits scattered throughout Uganda.[2] In late 1991, the Ugandan government announced the discovery of cobalt in Kasese; the deposits were about 1.1 million tons and could earn the country about $228 million.

Uganda lacks commercially exploitable oil reserves. In 1985, the World Bank funded the Petroleum Exploration Promotion Project, which sought to help the country define its policy toward oil exploration and development. As part of this program, geologists carried out field surveys in and around Lakes Albert and Edward. These activities, however, have failed to locate any petroleum deposits.[3]

Soils

Generally speaking, Uganda is blessed with fertile soil.[4] Indeed, those in the Lake Victoria zone and in the Mabira forests yield two crops a year, making them among the most productive in the world. There are, however, some significant variations between northern and southern soils, largely because of different rainfall and land use patterns.

In the south, the most common soil is red loam overlaying a compact clay subsoil. Southwestern soils are browner. These soils are deep, up to 15 inches, and extremely fertile. Similar soils are located in the foothills of the Ruwenzori Mountains and on Mount Elgon. Another good soil is the deep red loam found over Karagwe-Ankolean rocks in regions with a 50- to 70-inch mean annual rainfall. The topsoil, which can be 15 to 20 feet deep, is a dark chocolate color and is found in Bunyoro, Mubende, and eastern Kyagwe. In general, the north has poorer soils than the south. The third kind of soil, which is present on low-lying land, consists of a dark humic top layer from 3 to 12 inches thick on almost pure sand. In its extreme form it is nearly uncultivable. Such soil is found on the Lake George plain, in Gomba, east of the Masaka-Bukoba road, and in parts of the

Traditional housing (Photo courtesy of U.S. Committee for Refugees.)

Lake Kyoga drainage basin. An intermediate form of this soil is located over parts of eastern and northern Uganda. It has a thin top layer and the subsoil can be hard and gravelly.

Ironstone, which is also called murram, is present in all these soils. Normally, it is found as a nodular gravel or as solid ironstone somewhere between 1 to 40 inches below the surface. Murram makes excellent gravel roads. Solid ironstone, which is useful for building purposes, exists in large sheets in the central lake basin.

Soil erosion intensified after the introduction of cotton, which was cultivated on a million acres of new agricultural land. Cotton cultivation exposes the land to months of heavy rainfall and, until the plants mature, to intense sun. Large clearings for cotton cultivation result in the exposure of large pieces of land to soil wash. Since most farmers fail to construct cross drains, channels are created that carry away the best part of the soil. Despite this problem, erosion from runoff is minimal since most of the country is located on a plateau. Moreover, high rainfall speeds the growth of protective vegetation, which covers much of the south, and builds and restores soil fertility and structure.

Ever since colonial times, the government has sought to control soil erosion with a variety of reforestation and tree planting programs. Nevertheless, in 1968 forest reserves totaled only approximately 8 percent of the land, or about 5,770 square miles. By the early 1990s, the situation was worse. Theoretically, forests covered only 7.5 percent of the country's landmass; however, because encroachers had almost destroyed forests such as Kibale, Kisangi, Mount Elgon, Mabira, and Bukaleba, most foresters believed that the actual percentage was much lower.[5]

Climate

There are six climatic regions in Uganda: Lake Victoria, Karamoja, western, Acholi-Kyoga, Ankole-Buganda, and Mount Elgon. Although they usually merge near their edges, each region possesses some unique characteristics.[6]

The Lake Victoria region is about 30 to 50 miles wide. Because of adequate rainfall, bananas, coffee, and several other intensively cultivated crops grow well. This region is characterized by a group of flat-topped hills, which range in altitude from 4,300 to 4,400 feet; swampy inlets of the lake; and a swampy valley. At the lakeshore the daily temperature variation is about 13 degrees Fahrenheit; 30 to 50 miles further inland this increases to approximately 20 degrees. Although some rain falls every month, there are two dry seasons: from December to March and from June to July. Rainfall is greatest during the March–May and October–November periods. On the shores annual rainfall totals between 60 and 70 inches and occurs between 160 and 170 days per year. Elephant grass is the main vegetation in the southern parts of East Mengo and Mubende Districts, nearly all of West Mengo, and the eastern portion of Masaka District.

Most of Karamoja is on a wide, flat plain at an altitude of 3,500 to 4,000 feet. There are a few hill ranges with peaks rising to 8,000 to 9,000 feet. The absence of adequate rainfall and good soil requires the region's people to be pastoralists rather than farmers. With the exception of the west and northwest, much of Karamoja is semiarid. Rainfall is intermittent and averages about 15 inches in the dry regions to 35 inches on the western border. The rainy season is from April to August and the dry season is from November to March, when brushfires are common. During the dry season temperatures can be as high as 96 degrees Fahrenheit in the shade. In the rainy season high temperatures normally are about 85 degrees Fahrenheit in the shade.

The western region extends 30 to 70 miles inland from the Ugandan-Zairian border. It includes portions of West Nile, Bunyoro, Toro, Ankole, and Kigezi districts; and Lakes Albert, George, and Edward. The Ruwenzori Mountains and the Mufumbiro Volcanoes are in the southwest. In the northwest and southeast highlands, and the escarpment and hill east of Lake Albert, the elevation is anywhere between 4,000 and 5,000 feet. Temperatures can reach 90 degrees Fahrenheit or more throughout the region. Rainfall patterns are varied. Annually, the lake areas receive 34 to 40 inches, higher elevations get about 50 to 55 inches, and 60 to 80 inches fall on the slopes of the Ruwenzori Mountains. The dry season is from December to February. The rainy seasons are from September to October and from April to May.

The Acholi-Kyoga region includes parts of Acholi, West Nile, Madi, Teso, Lango, Busoga, and Bukedi districts. The area's altitude ranges from 3,000 to 4,000 feet. There are few hills. The most prominent geographic feature is Lake Kyoga. The papyrus swamp around the lake moderates the climate. Annual rainfall varies between 35 and 50 inches. The rainy season lasts from April to October. South of the swamp there is a dry season from December to March; north of the swamp there is low rainfall rather than a dry season during June and July.

The Ankole-Buganda region, which is on a plateau, comprises much of Busoga, Bukedi, and Ankole districts and most of the former kingdom of Buganda. There are hills in the north and south and swamps around Katonga River and Lakes Wamala and Kachira. There is an average of 40 inches of rain per year. Rainy seasons last from September to November and from April to May, and the dry seasons are between June and July and between December and February. Temperatures are moderate.

Mount Elgon, an extinct volcano, dominates this region, which includes parts of Bugisu and Sebei districts. In the south and southwest rainfall can amount to 70 inches a year; the northern slopes are somewhat drier.

Forests

Despite favorable climatic and soil conditions, Uganda has comparatively few closed (i.e., state-managed) forests.[7] In the 1930s, these forests covered only about 1,860 square miles, which put Uganda in a worse position than any other territory in the British Empire except Tanganyika (now Tanzania) and South and Western Australia. However, Uganda also possessed thousands of square miles of savanna forests, which were not under government control. Over the years, the latter gradually diminished in size, largely because farmers had cleared huge areas for crop production and grass fires had destroyed or damaged large tracts.[8] Despite this poor record, there are sizable forests throughout the country, especially in western Uganda, the Lake Victoria region. Smaller forests exist in many places, some of which include western Buganda, Ankole, Kigezi, West Nile, Budama, Bugwere, and Busoga.

These forests possess a wide array of species.[9] In Busoga's high grass savanna, for example, the mvule or Uganda iroko, with a bole that may rise to 70 feet and have a girth of 18 feet, dominates the landscape. The muwafu or incense tree resembles an oak. The kirundo or African upas has a slender white bole to 150 feet.

The mpewere, which can reach a height of 110 feet, produces red fruits the size of cherries. Other common trees include the African umbrella tree, the mahoganies, and the Uganda ironwood or muhimbi.

During the colonial period, the Department of Forestry, which was established in 1917, managed the country's forests to produce a sustained yield of timber, poles, and firewood. The Department of Forestry also maintained eucalyptus and cassia plantations near all the larger towns to provide the inhabitants with fuel and building poles. Additionally, the Department of Forestry planted hardwoods such as mahogany and mvule and established cypress and pine forests in the western highlands.

After independence, Uganda's forests were placed under increasing pressure by growing instability, an expanding population, and the collapse of effective government.[10] Agricultural encroachment, logging, charcoal making, and harvesting for firewood increased each year and caused widespread deforestation throughout the country. Neither natural regrowth nor tree-planting projects kept pace with the growing demand for forest products.[11]

After seizing power in January 1986, the National Resistance Movement regime promised to reverse these trends and to rehabilitate Uganda's forests. In 1988, the Ministry of Environment Protection assumed responsibility for administering forest policy. The following year, this ministry implemented a six-year Forestry Rehabilitation Project financed by the United Nations Development Program (UNDP) and the United Nations Food and Agriculture Organization. The project, which is expected to cost $37.2 million, included a nationwide tree planting campaign and a group of three-year training courses for farmers, rural extension agents, educators, and women's groups. As part of this project, the Department of Forestry devised a "set aside" policy whereby 50 percent of the country's high forests and savanna woodlands would be set aside as nature reserves and buffer zones. These areas would protect Uganda's forest cover, act as genetic pools, and offer refuges for wildlife.[12] Other forest programs included the Sustainable Development and Forest Conservation project and the Development Through Conservation project. The government also worked out a $6 million afforestation program, financed by Norway, to provide the inhabitants of Kampala, Jinja, Mbale, Arua, and Mbarara with firewood, building materials, and electric poles. Several Western nations, including the United Kingdom, Germany, and the United States, also provided financial or technical assistance to the forestry sector. On 7 April 1992, President Yoweri Museveni launched the National Tree Planting Programme, which eventually will require every Ugandan with a piece of land to use 10 percent of it as a woodlot to stop further environmental degradation.[13] He so outlined additional measures to preserve the country's forests, including requiring Ugandans to use cooking stoves that burn less charcoal and wood, biogas and solar power in homes and institutions, and charcoal briquettes made from coffee husks and other agricultural residue. Many ecologists maintain that unless these and other conservation measures halt the ever increasing destruction of Uganda's forests, some parts of the country could turn into a desert by the turn of the century.[14]

Game Reserves and Fisheries

Uganda has been home to an impressive array of wildlife. Some of the more popular species include elephant, lion, rhinoceros, leopard, topi, hartebeest, eland, and gorilla. In 1925, the colonial authorities established the Game Department to preserve and control the depredations of Uganda's fauna. By 1935, there were five game reserves in Uganda, including Bunyoro and Gulu (1,800 square miles); Toro (200 square miles); Lake George (266 square miles); Lake Edward (216 square miles); and Damba (12 square miles).

Increased poaching, especially of the elephant, and growing demand for land for economic development convinced the colonial authorities that Uganda's wildlife could be better protected by a national park system.[15] Unlike game reserves, national parks had territorial integrity that was inviolable and could be changed only by legislative authority rather than by mere administrative decision. In 1952, the Ugandan government therefore created the Queen Elizabeth National Park (764 square miles), in the region of Lakes Edward and George; and the Murchison Falls National Park (1,504 square miles), which stretched for about 1,000 miles along the Victoria Nile. For the next decade, these national parks succeeded in preserving an impressive array of Uganda's wildlife.

Unfortunately, this situation changed shortly after independence. Indeed, Uganda has compiled the most dismal and tragic game preservation record in postcolonial East Africa. Poaching initially presented the greatest threat to the country's fauna. However, after Idi Amin's military takeover in 1971, Uganda's wildlife suffered increasingly from the nation's decline into political, economic, and social chaos. Approximately 250,000 wild animals died at the hands of Amin's senior military officers, who poached for skins, ivory, rhinoceros horn, and meat. This carnage continued well into the next decade, largely because subsequent regimes were unable to restore stability. Rebel groups often killed elephants and sold their tusks, using the proceeds to buy arms and ammunition. In addition, many government officials engaged in the trophy trade for personal profit. Lastly, thousands of displaced persons and refugees relied on wild animals for food.

After seizing power in January 1986, Yoweri Museveni promised to end the slaughter of Uganda's wildlife. This proved to be difficult because the government allocated most of its resources to fighting rebels in northern, eastern, and western Uganda and to improving the country's economy. By the early 1990s, however, the Museveni regime had achieved some progress in conserving the country's fauna and restoring Uganda's tourist industry. On 8 May 1991, for example, the Ugandan government approved the establishment of the Ruwenzori Forest National Park and the Mgahinga Gorilla National Park.[16] Additionally, there are plans to turn the Bwindi Impenetrable Reserve and Animal Sanctuary and Mount Elgon Forest Reserve into national parks.[17] The tourist trade had achieved a 9 percent annual growth rate, with approximately 50,000 foreign tourists visiting Uganda in 1990.[18] Despite these encouraging signs, the rehabilitation of Uganda's national parks and tourism industry will be a slow process. Apart from continued instability in many areas, the country lacks an adequate tourist infrastructure. On 14 August 1991, the Ugandan government tried to rectify this situation by leasing Kyambura Game Reserve, which is the eastern buffer zone for Queen Elizabeth National Park, to Zwilling Safaris of Switzerland. The lease agreement stipulated that Zwilling Safaris would develop the reserve according to a plan approved by the Ministry of Tourism, Wildlife, and Antiquities. Considerable controversy over this arrangement emerged when journalists discovered that Zwilling Safaris had connections to South Africa.[19]

Poaching also continues to reduce Uganda's wildlife population. Wildlife authorities must cope with ill-disciplined NRA troops as well as with criminals. During a 21 October to 3 November 1991 military exercise in Mburo National Park, for example, soldiers from Kaburangire Army Training Wing killed impalas for food and used baboons for target practice.[20] In more recent years, the Ugandan government has succeeded in reducing poaching activities. This has allowed some wildlife populations, especially elephant, to increase. However, the government's lack of resources will hinder efforts to restore the country's wildlife herds for the foreseeable future.

Uganda possesses several lakes and rivers, which are stocked with nearly 200 species of fish. In the Nile system below Murchison Falls and in Lake Albert, the Nile perch, which can weigh up to 300 pounds, abounds. The upper waters of the Ruimi River, on the eastern slopes of Ruwenzori, carry brown trout, and the Suam and Bukwa rivers along the Ugandan-Kenyan border are stocked with rainbow trout. Other species common to Uganda include the tiger fish, barbels, and tilapia.

Commercial exploitation of these resources, which have been spared the destruction associated with much of Uganda's postindependence history, is one of the most significant sectors of the Ugandan economy. In the early 1990s, more than 9,000 Ugandans were engaged in small-scale fishing along the shore of Lake Victoria.[21] Fish are also an essential source of animal protein for Ugandans.

Despite its importance, the fishing sector has been threatened by several factors. Industrial waste from the Kilembe copper mines has polluted several rivers

The Nile perch, an essential source of protein for Ugandans. Given the growing environmental crisis in Lake Victoria, the future well-being of this species is by no means guaranteed. (Photo courtesy of U.S. Committee for Refugees.)

that flow into Lakes George and Edward and the Nile River.[22] Illegal harvesting of premature species and explosives fishing also endanger this valuable resource. In recent years, Lake Victoria fishermen have complained that they have nowhere to sell their catch because local industry cannot afford to buy or to process large amounts of fish. Kampala has also accused Kenyan fishermen of poaching in Uganda's share of Lake Victoria's waters. (Uganda's portion of the lake is about 40 percent; Kenya's share is only 5 percent.)[23] The lack of resources will prevent the Museveni regime from resolving these problems anytime soon.[24]

2

HISTORY

U GANDA'S HISTORY HAS BEEN DETERMINED largely by the divisions among its people. Linguistically, the country is divided between the Nilotic-speaking northerners and the Bantu-speaking southerners. Additionally, there is an economic rift between the pastoralists who live in northern and western Uganda and the agriculturalists who occupy much of the southern region. Political and territorial disagreements often caused tension and warfare between the ancient Ugandan kingdoms. Society also suffered from religious divisions as Islam, Roman Catholicism, and Protestantism competed for the loyalties of Ugandans.

The two most significant epochs in Uganda's preindependence history are the precolonial and colonial periods. During the first era, which lasted from the Early Stone Age to about the mid-nineteenth century, Bantu-speaking peoples who had migrated from western Africa settled throughout southern Uganda, and groups that spoke Nilotic and Sudanic languages inhabited the northern and eastern parts of the country. Some of the more important themes that characterized this period include the emergence of hunting and gathering societies; the subsequent development of food production, ironworking, and toolmaking; the rise of kingdoms; and the movement toward centralized state formation.

The arrival of European missionaries, explorers, and businessmen initiated a process that culminated in the 1894 declaration by Great Britain of a protectorate over Buganda, a kingdom whose agents then helped the British subjugate the entire country. The colonial period, which lasted until 1962, revolutionized Uganda. Apart from thrusting the country into the world economy, the British introduced Western medicine, education, law, administration, and government. Many Ugandans collaborated with the British in these efforts, but others resisted attempts to colonize their respective societies.

Precolonial Times

Details about the early precolonial period are sketchy and based largely on archeological findings and oral traditions. As early as the fourth century B.C., African cultivators, hunter-gatherers, and herders lived in the dense rain forest that surrounded the Lake Victoria basin. Intensive cultivation eventually destroyed almost all the region's tree cover. The cultivators who inhabited the forest seem to have been Bantu-speaking peoples who, apart from their agricultural expertise, used iron technology to make weapons and farming tools. It was these cultivators who gradually forced the hunter-gatherers to move their homes to more remote mountain sites. By the fourth century B.C., the Bantu-speaking peoples had perfected iron-smelting techniques that enabled them to produce medium-carbon steel in preheated forced-draft furnaces.

The rise of kingdoms and centralized states was a slow process that occurred over several centuries. Clan chiefs ruled the societies that gradually developed in the Lake Victoria basin. Eventually, as a result of the widespread cultivation of the banana, the Bantu-speaking peoples prospered and formed nascent states, a process that was accelerated by the arrival of Nilotic-speaking pastoralists who were en route from their traditional northern homelands in the Nile River area to grazing lands in central and southern Africa. Although these peoples lacked sophisticated political skills, they had the military strength to subjugate the agriculturalists. Eventually, four kingdoms (Bunyoro-Kitara, Buganda, Ankole, and Toro) emerged to regulate the political, economic, and military life of Uganda.

Bunyoro-Kitara, which was located in west-central Uganda, was the first centralized state to emerge in Uganda. The legendary Tembuzi people initially ruled this kingdom, whose inhabitants probably engaged in a variety of pastoral and agricultural pursuits. From about 1350 to 1500, the Chwezi people, pastoralists who had imposed themselves on the agricultural population, established a ruling dynasty in Bunyoro-Kitara. Among other things, the Chwezi introduced a hierarchical administration controlled by a king (*mulama*) who appointed senior palace officials and local chiefs.

After 1500, the more aggressive Bito, a Lwoo-speaking people from southeastern Sudan, replaced the Chwezi—who subsequently migrated to areas that would become Tanzania, Rwanda, and Burundi—as Bunyoro-Kitara's ruling dynasty. For the next several centuries, the Bito-ruled kingdom was the region's dominant political power.[1] Under the Bito kings, Bunyoro-Kitara established subdynasties in neighboring Bukoli, Bugwere, Bulamogi, Toro, and Bugabula in Busoga and in Kiziba in Tanzania's Bukoba District. Raiding parties ensured the loyalty of these subdynasties and harassed other principalities such as Buganda, Ankole, Rwanda, Karagwe, and Busoga.

Eventually, however, revolts in the outlying parts of the empire that resulted in part from the expansion of the Buganda kingdom caused Bunyoro-Kitara's links with subdynasties to weaken. Kabarega, Bunyoro-Kitara's king (*omukama*), tried to reestablish suzerainty over the outlying areas by creating new military regiments (*abarusura*). This strategy failed because secessionist factions in the subdynasties refused to recognize the king's authority. The emergence of Buganda—a smaller, more disciplined and efficient kingdom—accelerated Bunyoro-Kitara's decline as a regional power.

Buganda was a kingdom of immigrants. Some groups came from the northeast with Kintu, who ruled as Buganda's first king (*kabaka*) from 1395 to 1408. Other people migrated from Bunyoro-Kitara for personal or political reasons. Prince Kato Kimera led this group of refugees.[2] Eventually, Prince Kimera seized power and became Buganda's first strong king, ruling from 1447 to 1474.

Unlike Bunyoro-Kitara's *omukama*, Buganda's rulers allowed for political participation by all clans and people. To ensure continuity, they identified a new king with his mother's clan. When a *kabaka* died, a new king emerged from a group of eligible princes, each of whom belonged to his mother's clan. This ensured that a single clan could not occupy the throne for more than one reign, as would have happened if a king was identified with his father's clan.

During its early history, Buganda remained in the shadow of its more powerful neighbor, Bunyoro-Kitara. In Kabaka Katerega's reign (1636–1663), however, Buganda more than doubled its size by expanding its borders westward into the countries of Mawokota and Gomba, Butambala, and Singo.[3] By the early nineteenth century, Buganda had become the more powerful kingdom, and over the next several decades, it sought to destroy Bunyoro-Kitara's remaining influence.

By this time, Buganda had developed imperial tendencies. The *kabaka* appointed chiefs to rule newly conquered territories, dispatched royal embassies throughout eastern Africa, supported the creation of an internal trade system, and deployed his army to neighboring kingdoms and trouble spots throughout the kingdom. Buganda also developed a highly refined sense of its superiority, which caused a backlash among other peoples. As a result, Buganda increasingly relied on military force to control its empire. According to Henry M. Stanley, the American journalist who visited Buganda in 1875, the *kabaka* had organized a 125,000-man army and a fleet of 230 war canoes for a single campaign.[4] With the arrival of the British, Buganda's fortunes increased as the kingdom's officials became instruments of colonial rule.

Ruhinda, a leader of cattle-herding peoples, founded the kingdom of Ankole in a region known as Isingiro and became its first king (*mugabe*).[5] This kingdom consisted of two groups: Pastoralists who belonged to the Hima people formed the ruling group, and the Ira constituted the less privileged agricultural population. After Ruhinda died in about 1446, his empire disintegrated into small, independent kingdoms. As in Ankole, Hima pastoralists and Ira cultivators formed two classes. For at least the next fifty years, Ankole experienced considerable instability as dissident clans staged revolts against Ruhinda's son and his successor, Nkube.

By the seventeenth century, Ankole's expansionist neighbors, especially Bunyoro-Kitara, posed a greater threat to the kingdom than internal dissent. Nevertheless, Ankole managed to preserve its security without losing much territory. In the early eighteenth century, however, a military force from Bunyoro-Kitara invaded Ankole and forced the *mugabe*—Ntare IV, who ruled from about 1727 to approximately 1755—to abandon his kingdom. Ntare reclaimed his throne after inflicting a defeat on the Bunyoro army on its return from Rwanda. The *mugabe* further strengthened his position by instituting political reform and by reorganizing the army. As a result, Ankole succeeded in expanding its borders north to the Katonga River.[6]

The kingdom's growth continued for more than a hundred years, despite numerous succession struggles. The period of greatest expansion occurred during the reign of Mugabe Mutambuka (1839–1867), when raiding parties attacked neighboring states in the north and south. By the late nineteenth century, Ankole had reached the apex of its power; unfortunately, an outbreak of smallpox and a rinderpest epidemic weakened the kingdom and reduced its ability to resist the British.[7]

After learning that he would not succeed to the Bunyoro-Kitara throne, Prince Kaboyo established the kingdom of Toro in western Uganda in about 1830 and named himself king (*mukama*).[8] He then defeated a military force sent by his father, Nyamutukura, who wanted to destroy Toro. The region's people, who opposed what they perceived as Bunyoro-Kitara's imperialism and domination, welcomed Kaboyo's decision to create an independent kingdom.

Kaboyo's death, probably in the late 1860s, caused a series of secession wars that lasted until the 1870s. Bunyoro-Kitara, anxious to exploit this chaos, launched several military expeditions to regain control of Toro. Buganda, fearing a resurgence of Bunyoro-Kitara's influence, also competed for control of Toro. Machinations between the three kingdoms continued until 1891, when Captain Frederick (later Lord) Lugard, then in the employ of the Imperial British East Africa Company (IBEAC), signed a treaty with the kingdom of Toro. Apart from promising to protect Toro's independence, this document installed Kasagama as king of Toro. Kasagama was favorably disposed toward Buganda, having spent many years there in exile. This suited the British, who in 1890 had signed a similar treaty with Buganda that provided them with a secure base from which to extend colonial rule to the rest of Uganda.

The Advent of European Colonialism

Until the mid-nineteenth century, Uganda remained isolated from the outside world. However, coastal traders, searching for ivory and slaves, eventually established a settlement at Kafuro in the northern Karagwe kingdom. From this site, trade routes gradually extended into Buganda and Bunyoro and parts of northern Uganda. However, the most important factors that ended Uganda's isolation concerned the search for the source of the Nile River, which attracted an endless procession of European explorers seeking fame and fortune; Egypt's desire to extend its influence as far south as Uganda; and the activities of European missionaries.[9] In 1858, for example, Sir Richard Burton and John Hanning Speke[10] reached Lake Tanganyika only to discover that it was not part of the Nile River system. Speke then traveled alone to the northeast, where he became the first European to see a lake that the Baganda called Nnalubale. He named the lake after England's Queen Victoria and claimed that the body of water was the source of the Nile. Burton, who believed that the source of the Nile lay further to the east, later refuted Speke's discovery.[11]

Speke, hoping to preserve his reputation, returned to East Africa in 1860 with fellow explorer James Grant. The two visited Karagwe and Buganda. After leaving Buganda, Grant continued to Bunyoro while Speke traveled to Urondogani, forty miles downstream from Lake Victoria. Speke then walked upstream a few days until he came upon the source of the White Nile, a cataract that he named Ripon Falls. The two explorers eventually returned to England via the Nile Valley.[12] A "Nile duel" between Speke and Burton continued in the press and at Royal Geographical Society meetings until the former lost his life in a hunting accident on 15 September 1864.

After delivering a eulogy of Speke to the Royal Geographical Society, Sir Roderick Murchison indicated that he wanted to send David Livingstone to East Africa to examine the area between Lake Nyasa and Lake Tanganyika. Among other things, Livingstone was to discover whether any river flowed northward out of Lake Tanganyika to join the Nile River. Unfortunately, Livingstone failed to resolve the controversy surrounding Speke's explorations. In 1874, the *Daily Telegraph* and the *New York Herald* therefore sent Henry Morton Stanley to Lake Victoria to end the "Nile duel." Stanley confirmed not only that Lake Victoria was one body of water but also Speke's claim that Lake Victoria was the source of the Nile.[13]

Even before the Nile controversy had ended, Egyptian imperialism had become an important factor in opening Uganda to external influences. In 1869, Ismail, the khedive of Egypt, commissioned Sir Samuel Baker to command an expedition to the Great Lakes to suppress the slave trade and to bring the area under Egyptian administration. On 14 May 1872, Baker annexed Bunyoro to the khedive. However, Kabarega, the king of Bunyoro, resisted Baker's intrusion by launching an attack against his expedition. The Battle of Masindi, which occurred on 8 June 1872, forced Baker and his men to retreat to Foweira, a traders' station on the Nile about 60 miles northeast of Bunyoro. After his counterattack failed to defeat Kabarega's force, Baker concluded an alliance with Rionga, a traditional enemy of the Bunyoro king. This strategy not only failed to extend Egyptian influence into

Bunyoro but also determined Bunyoro's hostility toward Baker and his successor.[14]

In 1874, Colonel Charles Gordon succeeded Baker as governor-general of Equatoria Province. Gordon's orders directed him to suppress the slave trade, implement a monopoly over the ivory trade, levy taxes on the local population, create a network of military stations from Gondokoro to Buganda, annex Buganda, and launch steamers on Lake Albert and Lake Victoria. Shortly after his arrival at Gondokoro, Gordon dispatched two emissaries to Buganda in hopes of bringing Mutesa I's kingdom under Egyptian influence. However, Mutesa I resisted Gordon's machinations by diplomatic means, which included inviting European missionaries to Buganda and using the threat of force to persuade Gordon to abandon plans to build forts on Buganda's frontier.[15] Despite Mutesa I's resistance and limited resources, Gordon established an administrative system that laid the groundwork for Egyptian occupation of Equatoria for the next fifteen years. However, by the time Gordon left his post in 1876, he had failed to make any lasting inroads into Uganda.

In 1878, Emin Pasha, who had served on Gordon's staff, became governor-general of Equatoria. Unlike Gordon, Emin Pasha succeeded in establishing fairly harmonious relations with Bunyoro's king, Kabarega. However, he achieved little more than a state of armed neutrality with Buganda. Additionally, by 1880, Emin Pasha had reestablished posts that Gordon had evacuated on the Somerset Nile at Foweira and Foda. Over the next few years, Emin Pasha administered Equatoria competently and slowly gained the trust of many Africans. However, the Mahdist revolt, which all but destroyed Egyptian power in Sudan in 1883, stranded Emin Pasha in Equatoria. It was not until 1887 that a relief expedition, commanded by Henry Morton Stanley, rescued Emin Pasha and accompanied him to Bagamoyo, a town on the East African coast.[16] The isolation and eventual collapse of Emin Pasha's administrative infrastructure marked the end of Egypt's imperial designs on Uganda.

European missionary activity in Uganda began in 1877, when two Anglican missionaries who belonged to the Church Missionary Society (CMS) arrived in Buganda. Kabaka Mutesa, who feared the growth of foreign influence in his kingdom, restricted the missionaries to his court at Rubaga.[17] As a result, Christianity had little impact on Buganda's clan heads and general population.

In 1879, the first Roman Catholic missionaries, members of the Society of Missionaries of Africa (White Fathers), came to Kampala via Entebbe. Almost immediately, this French order started vying with the CMS for converts and influence. In 1883, the Catholic church authorized the establishment of the Vicariate Apostolic of Nyanza and Pope Leo XIII commissioned its care to the White Fathers.

The competition between the Protestant and Catholic missionaries alarmed Mutesa, who believed that England or France might intervene in Buganda to defend religious interests. He therefore played off one group against the other. After Mutesa's death in 1884, Mwanga became *kabaka* and continued his predecessor's policy of keeping the missionary orders divided. In 1885, for example, Mwanga allowed three Protestant converts to be killed and burned. A few months later, the *kabaka*'s men killed an Anglican missionary who had just entered Buganda. The greatest bloodletting occurred on 3 June 1886, when Mwanga authorized the burning to death of thirty-one Christians at Namugongo and the execution of many others at locations throughout his kingdom.

The major European powers coveted Uganda because it contained the source of the Nile River. After Great Britain seized Egypt in 1882, Uganda's strategic value increased because British officials believed that defense of Egypt required control of the Nile Valley, because damming or diverting the Nile's waters could ruin Egypt's agriculture. Although France, Belgium, and Italy threatened British interests on the Nile, Germany proved to be the greatest danger to Great Britain's imperial designs on Uganda.

Competing goals elsewhere in East Africa prompted London and Berlin to sign the 1886 Anglo-German Agreement. This agreement divided the areas between the Tana and Ruvumu rivers into British and German spheres of influence. However, the two countries failed to reach an accommodation on Uganda and southern Sudan, thereby laying the groundwork for a confrontation between London and Berlin.[18]

On 3 September 1888, the British Foreign Office (FO) granted a charter to the IBEAC to further Great Britain's interests throughout East Africa. As part of the IBEAC's strategy, a company agent, Frederick Jackson, agreed to reconnoiter the Lake Victoria region.[19] However, company officials warned Jackson not to become involved in Buganda's turmoil. On 2 February 1890, Carl Peters, who commanded the German Emin Pasha Expedition and was en route to Buganda, visited Jackson's base camp at Mumias in what is now western Kenya. After learning that Jackson was away on an elephant hunting expedition to Mount Elgon District, Peters intercepted a letter for Jackson from Kabaka Mwanga asking for help in thwarting Muslim attempts to drive him off the throne. The German adventurer then hastened to Buganda and concluded a treaty of friendship with Mwanga.[20]

But events in Europe prevented Buganda from falling under German influence. Great Britain and Germany, hoping to avoid a diplomatic confrontation, opened talks to resolve "outstanding disputes" between the two countries. On 1 July 1890, London and Berlin signed an agreement that ended Anglo-German competition in East Africa. According to the agreement's terms, Germany acknowledged the British protectorate over Zanzibar and Pemba Island; relinquished its protectorate over Witu and other coastal and inland territorial claims; and recognized a new boundary between British and German spheres of influence in East Africa westward to Lake Victoria and across it to the Congo Free State's boundary. In exchange, Great Britain ceded Heligoland Island in the North Sea and promised to encourage the sultan of Zanzibar to cede to Germany those mainland areas already leased to the German East Africa Company in return for an indemnity.[21]

The Anglo-German agreement marked the beginning of the British colonialization of Uganda. In late 1890, the IBEAC authorized Captain Frederick Lugard to undertake a diplomatic mission to Buganda to improve relations between the company and Mwanga.[22] On 26 December 1890, Lugard concluded a treaty with Buganda whereby the British obtained the right to intervene in the kingdom's internal affairs and assumed responsibility for maintaining internal order. Implementing these provisions proved to be a difficult undertaking. Religious conflict between Protestants, Catholics, Muslims, and pagans had polarized and destabilized Bugandan society and had revealed the kabaka's inability to rule his kingdom. Moreover, European missionaries competed with one another for converts and for influence at Mwanga's court. In May 1891, Lugard temporarily united the Protestants and Catholics by leading an expedition against Muslim forces along the Buganda-Bunyoro border. With the Muslim threat all but destroyed, Protestant-Catholic tensions increased to the point that both parties were on the verge of open conflict.

Rather than trying to defuse this potentially explosive situation, Lugard placed another IBEAC officer, Captain W. H. Williams, in charge of a small force at Kampala and departed Buganda on a seven-month tour of other African kingdoms. During his sojourn, Lugard signed a treaty of protection with the kingdom of Ankole, established a post at Katwe, enlisted the support of approximately 600 Sudanese troops, and restored Kasagama to the throne of Toro.[23] On the return march to Buganda, Lugard established a network of forts to safeguard Toro's independence.

After returning to Buganda in December 1891, Lugard learned that the IBEAC had issued instructions for him to withdraw from the kingdom because it lacked the funds to administer the territory. Lugard refused to obey this order and used moneys raised by the CMS in Great Britain to remain in Buganda until the end of 1892. Before leaving, however, Lugard altered the course of Buganda's history by supporting the Protestants against the Catholics during the Battle of Mengo.[24] Lugard then concluded a treaty with Kabaka Mwanga whereby the latter recognized the IBEAC's suzerainty over Buganda.

Having laid the foundation of British colonial rule in Uganda, Lugard returned to Great Britain and urged the British government to declare a protectorate over the country. Since the IBEAC did not have the resources to remain in Uganda, London sent Sir Gerald Portal, consul general of Zanzibar, on a fact-finding mission to Buganda.[25] Portal's report confirmed that the IBEAC was moribund. On 29 May 1893, Portal therefore concluded a new treaty of protection in the name of

Her Majesty's Government with Mwanga. As a result of Portal's activities and increasing domestic political pressure, the British government declared a protectorate over Uganda on 18 June 1894.

The British Colonial Period

Between the declaration of the protectorate and the outbreak of World War I, Great Britain consolidated its control over the Ugandan territories that had been in the IBEAC's domain. British colonial rule started in Buganda and then gradually spread to the rest of the country. During this process, Ugandans had to decide whether to collaborate with the British or resist the imposition of foreign rule.[26]

Among those who first collaborated with the British were the Baganda chiefs who accepted a political system in which discussions about the kingdom's affairs occurred in a council (*lukiko*). After 1895 the council met under the leadership of a British officer rather than under the *kabaka*'s authority. As the Uganda Protectorate grew, British colonial administrators increasingly relied on the *lukiko* model to control local politics and to disseminate principles of modern government.

The first person to resist the British was Kabaka Mwanga, who had become politically impotent because of the coalition between British officials and Baganda chiefs. In July 1897, Mwanga revolted and fled south to Buddu, where he tried to persuade his followers—mainly Muslims, pagans, and personal friends—to fight foreign domination. To eliminate this threat, the British launched a punitive expedition against Mwanga. After defeating Mwanga, the colonial administration deposed him and placed his infant son, Daudi Chwa, on the throne. Three Christian regents, all of whom were sympathetic to the British, ruled Buganda on Daudi Chwa's behalf.

Another threat to the British emerged in September 1897, when the Sudanese troops that had participated in the punitive expedition against Mwanga mutinied at Eldama Ravine (in present-day Kenya) because their pay was in arrears. Rather than joining an expedition to central Sudan, as ordered, the Sudanese soldiers returned to Busoga and joined another Sudanese garrison that already had imprisoned three British officers. A British force, commanded by Major J.R.L. Macdonald, pursued the mutineers and attacked them at Luba's Fort. After this engagement, the British colonial authorities who suspected the loyalty of the Sudanese garrisons in Bunyoro, Toro, Budda, and Kampala requested imperial troops from India.[27] The British, supported by Indian and Baganda soldiers, eventually suppressed the mutiny in early August 1898 by bombarding and then storming the last mutineer stronghold, a small stockade near Mruli.[28]

In 1900, the British formalized their relationship with the Baganda nobility by concluding the Uganda Agreement. (The British concluded similar agreements with Toro [1900]; Ankole [1901]; and Bunyoro [1933].) According to this agreement's terms, the Baganda acknowledged British sovereignty over their kingdom by agreeing to collect and pay taxes to the colonial administration. In exchange, the British preserved Buganda's traditional ruling hierarchy, which included the *kabaka, lukiko,* and *katikiro* (chief minister). However, the British retained ultimate authority over the activities of these offices. The 1900 Uganda Agreement also instituted a land tenure system that divided Buganda's land equally between the British crown and approximately 4,000 chiefs, most of whom held the land on the basis of individual private ownership. Lastly, the agreement gave Buganda a special status enjoyed by no other kingdom, which in turn promoted a feeling of superiority among the Baganda in the protectorate.[29]

From Protectorate to Republic:
The Early Period

The British conquest of Uganda was a slow process. The main reason for the gradual development of British colonial power concerned London's unwillingness to spend large amounts of money on military campaigns. Nevertheless, British administrators in Uganda eventually consolidated control over the entire country. The African response to this development was mixed, as some resisted the British while others collaborated with them.

Those who posed the greatest challenge to the British were leaders who fell afoul of the British. In 1893, for example, Mukama Kabarega attacked Buganda and Toro, both of which had signed defense treaties with the British. Kabarega resorted to military force because he feared, correctly, that the British had designs on Bunyoro-Kitara. The following year, the British, supported by Baganda troops, launched a punitive expedition against him. In February 1894, this force captured Kabarega's capital and cut his kingdom into two with a network of forts. However, Kabarega refused to surrender and continued efforts to drive the British from his kingdom. When these attempts failed, he and his followers crossed the Nile River into Acholi and Langi country. From these locations, he carried on a guerrilla war against the British and their Baganda allies.

Buganda's Kabaka Mwanga was another major resistance figure. In July 1897, he launched a revolt against the British largely because his power had been waning since he had accepted IBEAC protection. In retaliation, the British deployed a joint Sudanese-Baganda force against Mwanga, who then sought refuge in German East Africa. In early 1898, he left German East Africa, returned to Uganda, and joined Kabarega's guerrilla campaign against the British. On 9 April 1899, the British captured Mwanga and Kabarega and deported them first to Kismayo, Somalia, and then to the Seychelles Islands in the Indian Ocean.

One of the last large-scale attempts to resist British colonialism occurred during the 1907 Nyangire rebellion in Bunyoro.[30] Nyoro chiefs, who led the rebellion, had refused to cooperate with the British-appointed Baganda advisers and had opposed what they perceived to be Baganda imperialism. Although it used military force to keep the Baganda advisers in power, the colonial administration stopped appointing them in Nyoro.

African resistance ultimately failed because the British not only possessed superior weapons but also succeeded in implementing a highly successful divide-and-rule strategy. By encouraging and rewarding African collaborators, the British undoubtedly eased the difficult task of colonizing a country as diverse and politically complex as Uganda.

The Baganda were the most significant collaborators. Having established their dominance in Buganda as a result of the 1900 Uganda Agreement, the British used the kingdom's political system as a model for the country's other kingdoms. Over the next several years, the British spread their influence to Bunyoro, Toro, Ankole, Busoga, Gisu, Teso, Lango, Acholi, and numerous other lesser territories. Beginning in 1909, the British extended their authority to northeast Uganda by building a network of substations and stations at several important locations, including Kumi (1909); Palongo (1909); Gulu (1910); Kitgum (1912); and Arua (1914). In addition, the British created Kigezi District in southwest Uganda to check any possible Belgian or German claims on the area.[31]

During this process, the British often appointed Baganda advisers to local chiefs. This "subimperial" system enabled the British to rule Uganda by maintaining a network of African governments that owed allegiance only to the colonial administration.[32] However, many people throughout these areas resisted Bugandan political, cultural, and linguistic domination. In 1910 and 1911, there were numerous conflicts between British-appointed Baganda agents and local populations in Kigezi, Lango, Teso, and Bukedi districts. Despite this resistance, however, the British, with the help of their Baganda allies, gradually consolidated their rule and by 1914 controlled most of Uganda.

The Two World Wars

Although Uganda escaped the ravages of World War I, the conflict transformed the country's military establishment.[33] Elements of the 4th (Uganda) King's African Rifles Battalion participated in most of the East African campaign's battles. The colonial authorities enacted emergency legislation to upgrade this battalion to a regiment. Former members of the 4th (Uganda) King's African Rifles Battalion became part of a new Uganda regiment. By the end of World War I, Uganda had sustained the following casualties: killed in action and died of wounds, 225; died of disease, 1,164; and wounded, 760.[34] World War I exposed many Ugandans not only to the world beyond their traditional homelands but also to the vulnerability of their European rulers, many of whom died during the fighting.

During the interwar period, there were several issues that determined the course of Uganda's political development, including the land ownership dilemma in Buganda; the "lost counties" dispute between Buganda and Bunyoro; and the Closer Union debate. In the absence of any indigenous national-level political activity, most African activists focused their energies on trying to resolve these local problems.

As mentioned, the 1900 Uganda Agreement created a controversial land tenure system whereby the *kabaka* and approximately 4,000 of his chiefs received freehold rights to 9,003 lots of 1 square mile each known as *mailos*. This enabled the *kabaka* and his chiefs to establish clear title to some of the kingdom's best land and pass it on to their heirs. Additionally, the 1900 Uganda Agreement increased the chiefs' control over the peasant farmers who were tenants on their land and allowed the chiefs to grow prosperous, especially after the introduction of cash crops.[35]

In 1921, the clan heads (*batakas*), who were outside the ruling hierarchy, formed the Bataka Association to effect a revision of the land settlement. After Uganda's governor, Sir Robert Coryndon, refused to become involved in this controversy, the Bataka Association sent to the *kabaka* a list of *butaka* estates it claimed the chief had illegally seized. Although the *kabaka* agreed with the Bataka Association that the chiefs had wrongly acquired some *butaka* estates, he opposed the return of land to the *bataka*s. Instead, he urged that the *lukiko* pass a law whereby it would give a *mailo* estate to any *bataka* who proved that he had lost traditional land as a result of the 1900 Uganda Agreement. However, the *lukiko* rejected this recommendation and suggested that a more appropriate solution would be to give the *bataka* first option on the purchase of any *butaka* land that came up for sale.

Rather than accept this solution, the Bataka Association again appealed to the colonial government. After an official investigation, the provincial commissioner and the chief justice recommended the creation of a land arbitration court to return *bataka* lands to their previous owners. Although the governor accepted this recommendation, the British did nothing to facilitate the establishment of the land arbitration court. Instead, the colonial authorities expected the Buganda government to implement the report's findings. When the *lukiko* failed to act, Kampala submitted the matter to the British government for final adjudication. In October 1926, the secretary of state for the colonies, L. S. Amery, issued a report that, although critical of the *mailo* system, recognized the rights of the present owners to remain on their estates. Thus, Amery reaffirmed the special relationship that existed between the British and the kingdom of Buganda.[36]

The "lost counties" controversy began in December 1893, when the British declared war against Bunyoro. The colonial authorities had acted because they feared that Omukama Kabarega was determined to frustrate British interests. The British force that deployed against Bunyoro included many Baganda soldiers. After defeating Kabarega in February 1894, the British rewarded the Baganda for their loyalty by giving almost a quarter of Bunyoro's territory to Buganda.[37]

Recovery of these "lost counties" became a national obsession in Bunyoro.[38] In 1901, Kabarega's successor resigned as *omukama* because of human rights violations against his countrymen in the "lost counties." Then, in 1907, the Banyoro staged an armed uprising known as Ekyanyangire to force Baganda chiefs out of Bunyoro. Although many Baganda chiefs went into exile in Kenya as a result of this uprising, the "lost counties" remained under Buganda's control.

During the interwar period, the Banyoro escalated their campaign to regain the "lost counties." In 1921, disaffected Banyoro established the Mubende-Banyoro Committee, which petitioned the colonial government to restore the kingdom's territorial integrity. Every time Uganda's governor visited Bunyoro, dissidents petitioned him for the return of the "lost counties." Even after the British normalized relations with the kingdom by concluding the 1933 Bunyoro Agreement, the "lost counties" controversy remained a sensitive issue with many Banyoro officials and activists.

During the 1920s, Closer Union, which called for the establishment of an East African federation, caused considerable protest in Buganda. The Baganda ruling hierarchy, supported by many common people, opposed Closer Union because it feared that Kenyan settlers would dominate the federation and that it would lose its land to European settlement. Baganda leaders therefore launched an intense campaign to defeat Closer Union. In 1927, for example, the *kabaka* sent a letter to the secretary of state for the colonies warning that there was "no possible hope" of any benefit occurring to the Baganda as a result of Closer Union. Four years later, the *kabaka* unsuccessfully petitioned the Joint Select Committee on Closer Union to exclude Buganda from any federation. Serwano Kulubya, the primary Buganda delegate to the Joint Select Committee on Closer Union, also registered his opposition to the federation scheme by claiming that Closer Union would require Buganda to deal with a central government that would be largely influenced by the European population.[39]

Apart from these memoranda, the *kabaka* and his supporters placed great emphasis on the separate status that the 1900 Uganda Agreement had given Buganda. According to them, Buganda's uniqueness could not be altered to fit into a federal framework. Although agitation against federation dissipated after the Joint Select Committee on Closer Union issued a 1931 report opposing the scheme, the issue would emerge again in 1953, when there were renewed demands for federation between Kenya, Uganda, and Tanganyika (now Tanzania).

World War II again revolutionized Uganda's military forces. The colonial government recruited 77,131 Ugandans into nine infantry units, two field artillery batteries, and several auxiliary battalions. Ugandans served outside Africa for the first time, seeing action in Madagascar and Burma. In addition, Ugandans contributed to the defeat of the Italians in Abyssinia (now Ethiopia) and worked as part of a military labor force in Egypt and the Middle East. They also garrisoned at Mauritius and Diego Suarez near Madagascar and helped build defenses in Mombasa, Kenya. As in World War I, Ugandan soldiers received many awards, including the Distinguished Conduct Medal, the Military Medal, and the Member of the British Empire Medal.[40]

World War II also had a significant political impact on Uganda. An array of problems had alienated much of the protectorate's African population. Grievances included high prices for consumer goods, low commodity prices for farmers, low wages for workers, and an end to cheap Japanese textile imports. More important, the return of approximately 55,000 ex-servicemen, many of whom possessed political and organizational skills, helped mobilize discontent against the British.[41]

The Asian Community

Understanding the history of Uganda's Indian community requires an appreciation of India's long relationship with East Africa. Although Indian seamen and businessmen have traded throughout the Indian Ocean since at least 1500 A.D., it was only in the latter part of the nineteenth century that significant numbers of Indians started settling in East Africa. The region's earliest Indian inhabitants lived on Zanzibar Island and prospered under Sultan Seyyid Said's free trade policy. When the Sultan first arrived in Zanzibar, for example, there were 300 to 400 Indians on the island. By 1841 the Indian population had grown to about 1,000, and by 1866 there were approximately 6,000.[42] An ever increasing number of these traders eventually moved to the mainland to better control their businesses, which included trading in slaves and ivory.

Between 1896 and 1901, one of the first large contingents of Indians, which numbered 31,983, arrived in East Africa. They worked as indentured laborers on the Uganda Railway, which the British were building to link the seaport at Mombasa and the region's interior. Of those who worked on the railway, 25,259 returned to India, fell sick, or died; 6,724 chose to remain in East Africa. Some of those who stayed settled in Uganda. Over the next several decades, thousands of Indian immigrants, looking for economic opportunities, joined their compatriots in Uganda and became traders or businessmen.

The Indian community was not homogenous in origin, religious affiliation, or language. The majority of Indians who settled in Uganda came from Gujerat, Kathiawad, and Cutch on India's northwest coast. A smaller number were from the Punjab and Goa. About two-thirds of the Indians were Hindu who belonged to several castes, and the remainder were Muslims from the Shia or Sunni sects. The Indians spoke a variety of languages, including Gujerati, Punjabi, Urdu, and Hindi and to a lesser extent Sindhi, Maharatti, Konkani, and Bengali. Because they worked primarily in commercial enterprises, three-quarters of the Indians lived in towns scattered throughout Uganda's richest and most densely populated provinces (Buganda, Eastern, Western, and Northern).

Indian leaders formed several organizations to protect their interests. In 1908, for example, some businessmen formed the Kampala Indian Association to act as a focal point for the protectorate's Indians. The Central Council of Indian Associations (CCIA) in Uganda emerged in 1921. The council's goals included representing the Indian community; promoting unity; and protecting Indian political, economic, educational, and social interests. Three years later, a group of Indian leaders established the Indian Merchants' Chamber to influence the government's economic policy and Legislative Council deliberations.

The Indian community prospered under British colonial rule. Beginning in 1916, an increasing number of Indian businessmen became middlemen in the cotton industry. By 1925, Indians owned approximately two-thirds of Uganda's ginneries. Indian entrepreneurs also owned numerous sugarcane plantations and processing plants and operated a variety of businesses in the sisal, transport, lumber, tea, insurance, and retail industries.

Anglo-Indian political relations were uneven. Outwardly, there was little friction between the Indian and British communities and no significant anti-Indian discrimination in landholding or residential conditions.[43] However, the British did little to integrate Indians into Ugandan society, to discount the notion that Indians were only temporary residents in the protectorate, or to dispel the myth that Indians impeded African economic development. On a more practical level, Indians had many complaints against government policy.

When the first Legislative Council convened in 1921, for example, the governor appointed only one Indian as against two Europeans even though the Indian population was larger. As a result, Indian leaders boycotted the Legislative Council for five years. In the 1930s, the Indian community clashed with the government over several issues, including Indian opposition to a graduated poll tax, discriminatory marketing legislation, and British efforts to prevent an increase in Indian representation in the Legislative Council.[44]

At the outbreak of World War II, the Indian community escalated its attempts to influence policy. After the government enacted regulations for national service and started requisitioning private property, the CCIA formed a committee to investigate the political effectiveness of Kampala's Indians. The committee's report urged several reforms, including the adoption of a more representative CCIA constitution that would allow ordinary members to participate in the organization's activities. Additionally, the report recommended that the CCIA or the Kampala Indian Association, in consultation with the government, approve all

Indian nominations to public office. The uncertainty of the war years, coupled with continued Indian disunity, prevented the implementation of these and other suggested reforms.[45]

After 1945, a wave of Indian immigrants came to Uganda to share in the wealth brought about by the war. Indians also tightened their grip over Uganda's economy. Indian merchants, for example, controlled the wholesale field and owned most of the larger retail outlets. Two Indian sugar companies, Muljibhai Madhvani and Company, Ltd., and Mehta Sons (Africa) were the protectorate's main sources of private local capital. Additionally, both firms worked with the Uganda Development Corporation (UDC) to promote Uganda's agricultural and industrial development.[46]

This growing wealth exacerbated long-standing African resentment toward the Indian community.[47] In 1959, for example, the Uganda National Movement (UNM) boycotted Indian traders throughout Buganda. Using violence and intimidation, the UNM forced approximately half of the Indian businessmen out of Buganda. The British reacted to the boycott by arresting UNM leaders, banning the party, and threatening to terminate financial assistance to Buganda until the *kabaka* issued a statement condemning the violence.

Notwithstanding these drastic measures, the Indian community became increasingly insecure about its future in Uganda. Even before the boycott ended, Indians had taken unprecedented steps to improve relations with the Africans. In early 1959, a group of young radical Indians formed the Uganda Action Group (UAG), which argued that Indian security depended on the degree to which Indians identified themselves with African political aspirations. Despite its pro-nationalist orientation, the UAG failed to attract significant Indian or African support. As a result, its existence was short-lived; moreover, it was unable to narrow the gulf between the Indian and African communities.

As independence approached, Indian fear deepened. To protect Indian interests, conservative organizations such as the CCIA demanded constitutional safeguards such as separate electoral representation in the postindependence parliament. However, radical elements in the Indian community rejected this strategy, arguing instead that there be no racial representation in the postindependence parliament. Although the latter view eventually prevailed, Indian-African relations remained strained.

In particular, most Africans deeply resented Indians because they controlled much of the agriculture and commercial business activity in Uganda. Indians were also isolated, preferring to live among themselves rather than mixing with the various African communities. This inevitably led to charges of Indian racism. Many in the Indian community held contempt for the Africans, believing they were lazy and stupid. Because of the wide-ranging mutual antipathy between the two groups, the British failed to resolve the so-called Indian problem. As a result, this issue remained uppermost in the minds of most Ugandans after the country gained independence. Indeed, breaking what it perceived as an Indian strangle-hold on Uganda's economy became a major goal of the postindependence government.

The Colonial Economy

Uganda was the last East African country to be opened to the world economy. In 1901, the British completed construction of the Uganda Railway from Mombasa, Kenya, to Lake Victoria. Until then, ivory, Uganda's only export, had to be carried about 1,000 miles overland to the East African coast for onward shipment. The railroad enabled farmers and businessmen to transport their goods quickly and efficiently. As a result, Uganda enjoyed an excellent growth rate during the colonial period and by independence had one of the richest economies in sub-Saharan Africa. (Table 2.1 outlines the growth of the country's exports during the 1908–1965 period.[48])

For the first decade of British rule, Uganda's economy rested on peasant subsistence production, which was centered primarily in Buganda. With the exception of ivory, exports were negligible. The absence of significant economic activity and the limited use of cash restricted the colonial government's ability to collect taxes. As a result, the British Treasury had to provide an annual grant-in-aid to Uganda, which in 1903 was worth about 84 percent of government expenditure.[49]

By developing a viable export crop and moving toward a money economy, colonial officials hoped to end Uganda's reliance on the British taxpayer. Other factors that contributed to the impetus for economic development included the desire of traders and planters to boost profits and of Africans to improve their living standard. However, the most important reason was an increase in the world demand for raw materials, especially cotton.

TABLE 2.1 Total Value of Uganda's Exports, 1908–1965 (millions of pounds)

1908	140	1940	3,856
1910	212	1945	—
1915	507	1950	28,700
1920	1,828	1955	41,900
1925	5,097	1960	42,600
1930	2,188	1965	63,900
1935	3,631		

SOURCES: R.M.A. van Zwanenberg with Anne King, *An Economic History of Kenya and Uganda, 1800–1970* (London: Macmillan Press, 1977), p. 194; Jan Jelmet Jorgensen, *Uganda: A Modern History* (New York: St. Martin's Press, 1981), pp. 351–352.

The colonial government—supported by the Uganda Company (the commercial arm of the CMS), the British Cotton Growing Association, and the Baganda chiefs—therefore encouraged the development of an indigenous cotton industry. Starting in 1904, the colonial authorities, in conjunction with the Uganda Company, started distributing cottonseeds to growers in Buganda, Busoga, Bunyoro, and Ankole.

The protectorate's favorable climate and soil proved conducive to cotton cultivation. By 1907, cotton had become the largest item in Uganda's export trade. Moreover, between 1904 and 1910 there was a fourfold increase in revenue and an eightfold increase in the value of exports.[50] However, the crop was substandard, primarily because the protectorate lacked the infrastructure to produce quality cotton. In particular, the absence of an official training program forced growers to rely on their own skills, which often resulted in poor crops.

In 1908, the governor, Sir Hesketh Bell, introduced a protectorate-wide program to resolve this problem.[51] Among other things, he ordered the destruction of all cotton plants and the collection of all hand gins. To improve future cotton crops, Bell stipulated that Uganda would grow only one type of seed and that ginning would be done at central ginning stations. These reforms quickly improved the protectorate's cotton crop and facilitated the spread of cotton cultivation from Buganda to the Lake Kyoga region and parts of eastern Uganda. The increase in the number of growers resulted in greater harvests. Exports grew from £60,000 in 1909–1910 to more than £350,000 in 1914–1915. The resulting prosperity allowed the British government to reduce and eventually terminate its grant-in-aid to Uganda.

Apart from developing the cotton industry, the British supported other kinds of economic activity. Some officials supported the establishment of plantation agriculture to encourage the export of coffee, sugar, tea, tobacco, cocoa, and rubber. Officials favored such crops because they promoted the development of a plantation system of agriculture, which was controlled largely by Europeans and Asians. From the British point of view, the diversification of the agricultural sector ensured Uganda's success as a self-financing colony.

World War I crippled Uganda's fledgling economy. Many government departments experienced staffing shortages, often with disastrous economic implications. The few officers who remained on duty at the Veterinary Department, for example, failed to control a rinderpest epidemic that claimed 200,000 head of cattle. The war also nearly stopped the importation of capital for new enterprises. The decision to commandeer all but two steamers disrupted the lucrative Lake Victoria trade. Because the Uganda Railway was unable to cope with the extra traffic, businessmen had to store their goods for indefinite periods at various points along the railway. The end of hostilities did not improve Uganda's economic situation. Apart from the fact that it had neglected to make plans for demobilization and rehabilitation, the colonial government lacked the resources to provide relief to areas suffering from drought, famine, and food shortages.[52]

The war years also witnessed the emergence of a debate as to whether the colonial government should promote European-owned plantation agriculture or African cash crop agriculture. William Morris Carter, Uganda's chief justice and chairman of the Land Committee, advocated increased European settlement and greater support for European-owned plantations.[53] In 1915, Sir Frederick Jackson, governor from 1912 to 1918, approved Carter's recommendations and forwarded them to the Colonial Office in London, which had ultimate responsibility for Ugandan affairs. However, the following year the secretary of state for the colonies rejected the Carter scheme because it was not in the best interests of Uganda's African population. Instead, the British government eventually supported the development of African cash crops.

As a result, the Agriculture Department opened cotton research stations at Serere and Bukalasa and smaller experimental farms at several sites, including Ngetta, Bugusege, Kawanda, Mbale, Mukono, Bubulu, Wanyange, Kamuge, and Kyembogo.[54] The colonial administration also established the Bugisu Coffee Scheme to help African growers. This support helped to facilitate the growth of the cotton and coffee industries, which became the backbone of Uganda's economy. The colonial authorities regulated African production and marketing through mechanisms such as the 1926 Cotton Ordinance and the 1932 Native Produce Marketing Ordinance. African communities also formed local organizations to safeguard their interests. For example, the Buganda Growers' Society not only marketed cotton but also communicated the needs of African growers to the colonial government. Although export prices were subject to the pressures of the international marketplace, over which Uganda had little control, these two crops provided the basis for the protectorate's economic growth.

During the 1919–1939 period, an increasing number of African smallholding farmers throughout Uganda started growing one of these two crops. Eventually, about one-half of the protectorate's cotton cultivation occurred in Eastern Province, approximately one-quarter in Buganda, and a quarter in Northern and Western provinces. By 1930, cotton acreage was 304,000. Farmers grew robusta coffee in Buganda and near Lake Victoria and arabica coffee on Mount Elgon's slopes in Eastern Province. By late 1934, there were 17,816 acres under arabica cultivation and 26,224 acres under robusta cultivation.

Apart from cotton and coffee, Uganda produced several other export crops, the most important of which were sugar and tea. In the 1920s, Indian entrepreneurs established sugarcane plantations and processing plants near Jinja, where there was adequate rainfall. After 1924, European and Indian planters took a greater interest in growing tea, which was confined largely to Western Province, particularly Toro District, and Buganda. By late 1934, there were 1,290 acres under tea cultivation.

Coffee plantation (Photo by Thomas P. Ofcansky.)

After World War II, Uganda experienced considerable economic growth.[55] Cotton and coffee continued to dominate the protectorate's exports though coffee replaced cotton as Uganda's leading export. However, the protectorate's dependence on these two crops, both of which were subject to world price and domestic output fluctuations, became a major economic weakness. The British sought to resolve this problem by diversifying Uganda's economic base. Apart from encouraging the expansion of the sugar, tea, tobacco, and livestock industries, the colonial government committed itself to the more difficult task of developing the industrial sector.

Sir John Hall, who was Uganda's governor from 1944 to 1952, was a strong advocate of industrialization. In Hall's view, agriculture was sufficient to satisfy Uganda's immediate needs; however, given the protectorate's rapidly expanding population, only industrialization could support future generations.[56] In response to Hall's vision, the colonial government formulated a multifaceted scheme to industrialize Uganda.

Central to this plan was a hydroelectric dam on the Nile River at Owen Falls near Jinja. A factory at Tororo was to provide the cement for this installation. Additionally, Kampala devised a project to extend the railway to Kasese to provide a railhead for the Kilembe copper mine. Ore from this mine would then be transported to a smelter, which would be operated with cheap electricity, at Jinja. Electric power would also be used to run a textile factory at Jinja. Although claims that the dam would transform Jinja into the "Detroit of Africa" proved to be overly optimistic, the British succeeded in building the dam, the cement and textile factories, and the railway extension to Kasese.[57]

In 1952, the colonial government established the Uganda Development Corporation to facilitate further the protectorate's industrial development. Although it failed to promote small-scale industries, the UDC assisted in the development of several industrial projects. In 1956, for example, an asbestos plant opened in Tororo, and a cotton mill started working in Jinja. The UDC also helped operations at the Kilembe copper mine, which exported more than US$2.8 million of blister copper in 1957 and more than US$8.4 million worth in 1960.[58]

By the end of the colonial period, Uganda was still largely dependent on peasant agriculture, despite British efforts to industrialize and modernize the economy. Cotton and coffee remained a major source of income, accounting for more than 80 percent of Uganda's export earnings. However, other cash crops such as sugar, sisal, groundnuts, tobacco, tea, and livestock had also become important components of the protectorate's economy. By 1960, for example, earnings from these crops totaled about US$70 million, the same amount generated by coffee and cotton. Uganda's economic situation compared favorably with that of its two sister colonies, Kenya and Tanganyika. The protectorate's annual per capita gross domestic product was approximately US$64.4 million, which was higher than Tanganyika's but lower than Kenya's. The economic task that faced independent Uganda was the same as that which had confronted the colonial government; namely, how to accelerate economic growth and improve the standard of living of its citizens.[59]

In the years approaching independence, it appeared that postcolonial Uganda would be able to achieve these goals with ease. The agriculture sector was strong. A growing manufacturing sector appeared capable of making significant contributions to the country's economic well being, especially through the production of foodstuffs and textiles. The question of the status of the Asian community, which controlled much of the Ugandan economy, remained in abeyance as the colonial authorities, through their inaction, left the problem for the independent Ugandan government.

Growing Discontent in Buganda and the Kabaka Crisis

Nationalist stirrings emerged after World War II as evidenced by the formation of several political parties that helped prepare the way for Uganda's independence. However, unlike many of Great Britain's African colonies, mass nationalism never became a significant factor in the struggle for independence, largely because of the emergence of Bugandan separatism. The prospect that Buganda would not be part of a united and self-governing Uganda alarmed the British and Ugandans alike. A disintegration of Uganda would have far-reaching economic conse-quences insofar as Buganda was the most populous and wealthy province of the protectorate. An independent Buganda would also encourage separatism among Uganda's other ethnic groups. It was in this uncertain atmosphere that Uganda gradually approached independence.

Even before World War II had ended, discontent in Buganda signaled a growing dissatisfaction with the colonial order. Several political and economic grievances kindled popular opposition to the chiefs, who had prospered under colonialism, and to the British. In particular, many Baganda opposed Closer Union with Kenya and Tanganyika because they feared that Buganda would be forced to join an East African federation that would be dominated by Kenya's European settlers. Also, an increasing number of young, educated Baganda resented the political and eco-nomic advantages enjoyed by the chiefs and demanded greater democratization of Buganda's institutions. Lastly, the Baganda distrusted the colonial govern-ment's marketing and pricing policies, which they believed discriminated against the small farmer.

To better articulate these grievances, the Baganda established the Bataka Party in 1945 and the Uganda African Farmers' Union (UAFU) two years later. These groups quickly attracted widespread village-level support for their hostility to-ward the chiefs and the British colonial government. In 1949, the Bataka Party, supported by the UAFU, presented a memorandum to the *kabaka*. This document demanded sixty unofficial (non-chief) members to the *lukiko*, popular election of chiefs, and the resignation of the Bugandan government. After outbreaks of vio-lence in April and May and again in September 1949, the authorities arrested about 1,300 people, including the leaders of the Bataka Party and the UAFU, and banned both organizations.[60]

Although it reflected a high degree of political consciousness among the Baganda, this unrest was local rather than national in orientation. As a result, the 1949 incidents supported the notion that the Baganda considered their political destiny apart from that of the country as a whole. This sense of separateness also explained Buganda's opposition to the colonial government's 1949 decision to have two Baganda representatives on the Legislative Council. On a broader level, Buganda worried that independence would end its special status guaranteed by the 1900 Uganda Agreement.

In January 1952, developments in Buganda took an unexpected turn when Sir Andrew Cohen assumed the governorship of Uganda. Unlike his predecessors, who saw Ugandan politics through the lens of Buganda, Cohen maintained that Buganda should remain a province in a united Uganda. He also advocated greater democracy in Buganda's institutions. To achieve these goals, Cohen implemented a series of political and economic reforms. Among other things, he persuaded the *kabaka* to appoint three more ministers to his government and to accept a procedure whereby the *lukiko* would approve all future ministerial appointments. Additionally, Cohen supported a policy that increased the number of elected members in the *lukiko* to sixty out of eighty-nine.

Although these actions alarmed many Baganda politicians, it was the so-called *kabaka* crisis that opened a gulf between Buganda and the colonial government. The factors that caused this crisis included Buganda's rejection of the East African Federation; the kabaka's demand that the British government transfer Buganda from the Colonial Office to the Foreign Office; and the *kabaka*'s insistence that the British fix a timetable for Buganda's independence separate from the rest of the protectorate.

In October 1953, Governor Cohen held a series of meetings with the *kabaka* to resolve these problems. When the two sides failed to reach an agreement, Cohen ordered the *kabaka*'s deportation to England. Although most Baganda had perceived Mutesa as little more than a British puppet, Cohen's action made him a national hero. Over the next several months, there were increasing demands by Baganda from all walks of life for the return of their leader.

The Colonial Office then asked Sir Keith Hancock, director of the Institute of Commonwealth Studies, to devise a solution to the *kabaka* imbroglio. Hancock discussed the matter with the Buganda government's delegates at the Namirembe conference. In 1955, the two sides signed a new Uganda Agreement. Under the agreement's terms, Mutesa became a constitutional monarch and the *lukiko* accepted responsibility for electing the *kabaka*'s ministers, subject to the governor's approval. Buganda also abandoned demands for a separate independence. In October 1955, Mutesa returned to Buganda and received a hero's welcome. Although many British officials maintained that the agreement had weakened the *kabaka*'s power, the fact of the matter was that Mutesa had gained the right to appoint members of his government. Additionally, the agreement strengthened Buganda's position in Uganda. As a result, there was considerable tension between Buganda and the protectorate's other kingdoms during the last years of colonial rule.

The Rise of National Political Parties

Concurrent with the colonial government's difficulties with Buganda was the emergence of nationally oriented political parties.[61] Although these parties did little to improve the lives of average Ugandans, they succeeded in laying the groundwork for Uganda's independence. The proliferation of political parties also reinforced the ethnic divisions that continue to plague Uganda. Additionally, these parties introduced the concept of national politics to Buganda, which gradually involved more and more Baganda politicians. From the colonial government's point of view, this was a welcome development insofar as it helped to discredit the notion of Bugandan separatism.

In 1952, Ignatius Musazi and Abu Mayanja formed Uganda's first political party, the Uganda National Congress (UNC), which represented the Protestant educated elite from Buganda and other parts of Uganda. This party supported national unity, self-government, universal suffrage, and the Africanization of Uganda's economy. From its inception, the UNC suffered from many shortcomings that undercut its political credibility. Apart from the fact that its leadership was restricted to Baganda Protestants, the UNC lacked a national program. Musazi and his followers devoted their energies to exploiting local issues on a district-to-district basis. This strategy exacerbated rather than healed Uganda's ethnic divisions. As a result, dissident elements emerged within the UNC and argued for the formation of a new, more nationalistic political party.

In July 1957, a group of intellectuals who opposed Musazi therefore renounced their UNC membership and established the United Congress Party (UCP). In 1959, another split occurred when some of Musazi's supporters revolted and created a rival party, initially known as Uganda Nationalist Movement, led by Milton Obote. Infighting within and between the Musazi and Obote wings eventually destroyed the UNC.

The Progressive Party (PP), founded in 1955, enjoyed its greatest support among conservative Baganda intellectuals, landlords, and businessmen. E.M.K. Mulira was the party's first president. Unlike the UNC, the PP formulated a national platform, which called for a federal constitution and for the *kabaka* to become head of state. However, the party lacked the leadership and popular support to achieve its goals. Instead, elitist factions battled one another for control of the PP, until Mulira resigned as president and the party collapsed in disarray.

The Democratic Party (DP), established in 1956 by Matayo Mugwanya, appealed largely to Baganda Catholics who opposed Protestant control of the appointment process for public office and who feared UNC-inspired Protestant domination of Uganda. The DP supported anticommunism, Africanization of the civil service, and a unitary independent Uganda. After failing to conclude an alliance with the PP, its Protestant counterpart, the DP launched a recruitment campaign to enhance its popular support. Apart from activities in Catholic-controlled areas, the DP targeted Catholic Banyaruanda migrant workers in Buganda and the Bunyoro in the so-called lost counties. The party's recruitment efforts enjoyed considerable success in Kigezi, Ankole, West Nile, Bunyoro, and eastern Acholi. The increased membership enabled the DP to maintain a continuous existence from 1956 to 1969 and to play a major role in the decolonization process.

In 1958, George Magezi and William Nadiope, two Legislative Council members, created the Uganda People's Union (UPU). This party, which had no formal organization outside the Legislative Council, was unique insofar as it was not

Buganda-based or Baganda-led. Many non-Baganda, especially those from the western kingdoms and Busoga, therefore perceived the UPU as a bulwark against Bugandan imperialism. As the only pre-1960 nationwide political party, the UPU claimed that it represented Uganda's only genuine nationalist movement. The protectorate's other political parties, all of which were based on narrow ethnic and religious interests, were unable to refute the UPU.

A more powerful broad-based party emerged in 1960, when the UPU and Obote's wing of the UNC merged to form the Uganda People's Congress (UPC). Obote became party leader. His strategy, embodied in the slogan "Unity-Justice-Independence," called for consolidating party politics, acquiring immediate independence, and creating a strong postindependence central government.

The growth of nationally oriented political parties revealed many of the weaknesses that would plague postindependence Uganda. Chief among these was the inability of any politician to establish a significant base of support among peoples of different ethnic and religious backgrounds. That Baganda politicians dominated many parties was indicative of an inescapable fact of political life in postindependence Uganda: namely, that no government would enjoy stability without Buganda's support.

Decolonization and Independence

During the 1960–1962 period, divisive forces threatened to give birth to an ethnically divided independent Uganda. In mid-1960, the *kabaka* asked the British government to postpone the 1961 Legislative Council elections until the two sides had completed arrangements for a federal relationship between Buganda and the rest of the protectorate. Bunyoro, Toro, and Nkore made similar demands on London. After the British government decided to proceed with the elections, the *lukiko* boycotted them and on 31 December 1960 issued a unilateral declaration of independence. Although Buganda never seceded from the protectorate, the incident reflected the kingdom's desire to maintain its special status in a self-governing Uganda.[62]

The DP and the UPC contested the 1961 elections. The few Baganda who defied their leaders and registered to vote voted for the DP, which won 20 of the 21 Buganda seats and 43 seats overall. The UPC won 35 seats, the UNC 1 seat, and the independents 2 seats. After the DP victory, the governor, Sir Andrew Cohen, appointed DP leader Benedicto Kiwanuka head of the Legislative Council and minister without portfolio.

In July 1961, the Relationships Commission (Munster Commission), which the colonial government created to determine the future relationship between the protectorate's kingdoms and the central government, released its report. The commission recommended the establishment of a strong central government with a federal relationship for Buganda and semifederal status for Bunyoro, Toro, and Nkore. Additionally, the commission called for the direct election of Buganda's *lukiko* members and conferred upon that body the option of selecting National Assembly representatives by direct or indirect elections. The DP rejected the commission's findings, largely because it would be at a disadvantage, given its unpopularity with Buganda's leaders, should the *lukiko* refuse to hold direct elections.[63]

The 1961 Uganda Constitution Conference, held in London, ended the Relationships Commission controversy by laying the groundwork for the protectorate's postindependence government.[64] Accordingly, the Internal Self-Government Constitution, which took effect on 1 March 1962, transformed the Legislative Council into a unicameral National Assembly of eighty-two directly elected members. Additionally, the conference stipulated that elections be on a common roll with universal adult suffrage. The governor retained responsibility for foreign affairs and defense. However, Prime Minister Kiwanuka and a cabinet, responsible to the National Assembly, controlled all other government functions. As for Buganda, the conference concurred with the Relationships Commission's suggestion that the kingdom have a federal relationship with the central government. Moreover, the conference recognized Buganda's right to decide whether to hold direct or indirect elections for its assembly.

Even before the constitutional conference had ended the opposition, UPC and its traditionalist allies in the Buganda government started planning their strategy to defeat the DP in the *lukiko* elections and the general elections, scheduled for February and April 1962 respectively. The traditionalists considered but rejected the idea of joining the UPC, which had little popular support. Instead, they established a movement called the Kabaka Yekka (The King Alone) to contest the *lukiko* elections.[65] The movement subsequently won 65 seats whereas the DP gained only 3 seats. The *lukiko* then decided to elect Buganda's representatives to the National Assembly. Without direct elections, the DP had no chance of winning any National Assembly seats.

In the general election, the DP therefore campaigned from a position of weakness. Nevertheless, Kiwanuka enjoyed considerable support in Toro, Bunyoro, and Nkore, all of which feared domination by the *kabaka*. As a result, the DP won all but 2 of the seats in the three kingdoms. Overall, however, the DP won only 24 seats whereas the UPC gained 37 seats. The lack of a majority prompted the UPC to form a ruling coalition with the Kabaka Yekka under Milton Obote's leadership. This government led Uganda to independence on 9 October 1962.

3

GOVERNMENT AND POLITICS, 1962–1994

THE DREAM OF UGANDAN INDEPENDENCE quickly became a nightmare from which the country has yet to emerge. Understanding the dimensions of this tragedy requires an appreciation of the interrelationship between Uganda's ethnic diversity, the central government's increasing ineffectiveness, the emergence of the military as a political actor, and the proliferation of weak, brutal, and incompetent leaders. The current regime, though committed to restoring Uganda's stability and ending the violence that has characterized the postindependence period, still has yet to come to terms with these factors. As a result, Uganda remains a divided society that is unable to provide for the political, economic, and social well-being of its people.

Obote's First Government

Uganda's first postindependence government consisted of a coalition between Prime Minister Obote's Uganda People's Congress (UPC) and Buganda's traditionally oriented Kabaka Yekka (KY). Conciliation between the central government and Buganda occupied most of Obote's first months in office.[1] To placate Buganda, Obote assigned four cabinet posts to KY members of the National Assembly, accepted Buganda's federal status, and married a woman from Buganda. In 1963, the prime minister made another conciliatory gesture to Buganda by making the *kabaka* president after Uganda became a republic. Although this office was ceremonial, the appointment was the apex of UPC-KY cooperation.

Obote's ruling coalition eventually broke up over the "lost counties," which the British had assigned to Buganda in 1900. Rather than risk damaging the negotiating process at the 1962 constitutional conference, the British postponed a referendum on the "lost counties" matter until after Uganda had achieved independence. Although Buganda repeatedly voiced its opposition to a referendum, Obote scheduled a vote for November 1964. Residents of the "lost counties" had to choose between remaining part of Buganda, returning to Bunyoro, or becoming a separate district under the central government's control.

To improve its chances of winning the referendum, the Buganda government launched a soldiers' settlement scheme in Buyaga County. This strategy failed because of graft and corruption associated with the scheme and the central government's decision to restrict voting eligibility in the referendum to those who had lived in the "lost counties" before 1962.[2] As expected, the largely Bunyoro electorate voted to return to Bunyoro. The vote was 13,600 for returning to Bunyoro, 3,500 for remaining part of Buganda, and about 100 for falling under the administration of the central government.

Buganda protested the referendum's results and Kabaka II unsuccessfully petitioned the courts to block the transfer of the "lost counties" to Bunyoro. Notwithstanding the link between Obote and the "lost counties" controversy, UPC influence grew during the postreferendum period. By the end of 1964, several DP and KY members of parliament had quit their parties and had joined the UPC. The following year, the KY disbanded and urged its followers to enlist in the UPC. Many Baganda believed this action would put them in a better position to influence government policy. Despite its dominant position in Ugandan politics, the UPC experienced a series of conflicts that eroded its power and cast doubt on its ability to rule the country.

At the 1964 UPC party conference in Gulu, Grace Ibingira, a cabinet minister from Ankole, defeated Obote confidant John Kakonge for the post of secretary-general.[3] Ibingira quickly gained the support of anti-Obote elements and soon controlled about one-third of the cabinet. After purging the UPC of leftists, Ibingira plotted to launch a palace coup to oust Obote. On 4 February 1966, Ibingira called a cabinet meeting while Obote was on an up-country tour of Acholi and West Nile. Only nine of eighteen cabinet members attended the meeting. Daudi Ochieng accused Deputy Army Commander Idi Amin—in collusion with Obote and two cabinet ministers, Adoko Nekyon and Felix Onama—of involvement in a gold and ivory scandal.[4] Ochieng then moved a motion demanding Amin's suspension from the army pending an investigation of the matter. The motion carried, with the only dissenting vote being cast by John Kakonge.

Obote returned to Kampala on 12 February 1966 and denied the charges against him. He then went on the offensive against his critics. During a 22 February 1966 cabinet meeting, Obote arrested Ibingira and four other ministers (Emmanuel Lumu, George Magezi, Balaki Kirya, and Mathias Ngobi). The following day, Amin, who was regarded as an Obote loyalist, received a promotion to army chief of state. Obote then suspended the 1962 Constitution, which had recognized Buganda's privileged position, and removed the *kabaka* from the presidency. In April 1966, he convened the National Assembly, which ratified a republican constitution. The National Assembly also created an executive presidency for Obote, reduced the power of Uganda's traditional leaders, and ordered direct elections for all government offices. The Supreme Court upheld Obote's so-called revolution.[5]

The resulting gulf between the central government and Buganda caused a serious political crisis. In May 1966, Buganda's *lukiko* rejected the new constitution and ordered the central government out of the kingdom. Additionally, the *kabaka* asked for foreign military aid and prepared to launch an insurrection. Obote responded to this threat by declaring a state of emergency and by ordering army units to attack the *kabaka*'s palace on Mengo Hill. Although Mutesa II escaped to exile in London, the military crackdown, which resulted in more than 100 deaths, crushed Bugandan separatism.[6]

With most of his strong opponents eliminated, Obote moved to consolidate his power. In 1967, he introduced a new constitution, which abolished Uganda's four kingdoms (Buganda, Bunyoro, Ankole, and Toro) and gave the president considerable powers. Obote also purged the UPC of members suspected of being loyal to Ibingira or the banned KY. More important, he launched a "Move to the Left" program to create a socialist-oriented society.[7]

During the 1969–1970 period, Obote issued five documents to implement this program. The first, known as the "Proposals for National Service," called for every Ugandan to serve one or two years of compulsory national service. Document No. 2, "The Common Man's Charter," rejected feudalism and tribalism. Moreover, it promised to improve the standard of living for all Ugandans, establish agricultural cooperatives, and nationalize some industries and land to enable the Ugandan people to control the means of production. On 19 December 1969, delegates attending a UPC conference approved the charter and the "Move to the Left" program. Following an address to the delegates, Obote survived an assassination attempt. The cabinet, which was filled with Obote supporters, used the attack to justify the banning of all opposition groups and the creation of a one-party state. The third document, entitled "His Excellency the President's Communication from the Chair of the National Assembly on 20th April 1970," unified salaries for state employees and eliminated some of their prerogatives such as low-interest loans and overtime pay. Document No. 4, the "Nakivubo Pronouncements," outlawed strikes and demanded the 60 percent nationalization of eighty large commercial firms. The last document, entitled "Proposals for New Methods of Election of Representatives of the People to Parliament," required every National Assembly candidate to stand for election in three regional constituencies as well as the home constituency. Obote believed this "one-plus-three proposal" would reduce ethnic conflict and encourage loyalty to national institutions.[8]

Obote's "Move to the Left" program never reached fruition. However, his downfall came about from his increasingly poor relations with Amin rather than his determination to transform Uganda into a socialist society. Although Amin had prospered under the Obote regime, the two drifted apart after Amin had become commander of the army. Amin established control over the armed forces by eliminating rivals, promoting officers loyal to him, and recruiting mercenaries from former Anya Nya guerrillas in southern Sudan. During this process, many Ugandans came to doubt Amin's reputation as a capable military commander. Obote accused him of misappropriating money that the government had allocated to the military. Other government officials suspected Amin of arranging the 1970 murder of his sternest critic in the army, Brigadier Pierino Yere Okoya. To counter Amin's influence in the armed forces, Obote moved some of his own supporters into senior positions. In addition, he seconded pro-Amin officers into the civil service. On 25 January 1971, Amin responded to these challenges by overthrowing Obote while he was in Singapore for a Commonwealth Summit Conference.[9]

The Amin Years

Idi Amin destroyed Uganda. Understanding this tragedy requires an appreciation of his personality and his behavioral deficiencies. In about 1925, Amin was born in a small village near Koboko in West Nile Province. He is a Kakwa, a minority group whose origins are Sudanic-Nubian. In Uganda, any peoples from West Nile are called Nubians. One author has claimed that this group is renowned for its "sadistic brutality, lack of formal education, for poisoning enemies and for their refusal to integrate."[10]

Amin's formative years were spent as a noncommissioned officer and a sergeant-major in the King's African Rifles (KAR). According to Major Iain Grahame, one of his commanding officers, Amin was an "outstanding soldier" and a "born leader of men." Grahame also observed that these qualities coupled with Amin's "limited intelligence" created "a potentially explosive combination."[11] After becoming president, Amin's personality underwent the transformation alluded to by Grahame. His innate paranoia caused him to unleash a reign of terror against anyone suspected of disloyalty or treachery.

Amin's savagery fell into two periods. The first occurred during the initial year of his presidency when Amin feared the influence of the Acholi and Langi, groups that dominated the armed forces. The second started after Milton Obote launched an unsuccessful coup attempt on 17 September 1972 and lasted until the demise of the Amin regime.

*Idi Amin portrayed as an African Hitler. (Photo by
Thomas P. Ofcansky.)*

The tactics Amin used against his adversaries defy description. He did not merely order his henchmen to kill people but also encouraged them to subject their victims to unspeakable atrocities after they were dead. Ugandans regularly discovered the mutilated bodies of their relatives, friends, and acquaintances with "livers, noses, lips, genitals or eyes missing." Amin himself was suspected of cannibalism and practicing traditional Kakwa blood rituals on slain enemies.[12] Unlike the brutal methods used by many tyrannical governments to achieve political goals, the violent tactics employed by the Amin regime often had no purpose other than to terrorize the Ugandan population.

Sadly, few in Uganda suspected the terrible fate that was in store for their country when Amin seized power. The day after the coup, for example, there was widespread celebrating in Buganda and Busoga, where there had been considerable discontent with Obote. Some of Uganda's more prominent personalities—including Grace Ibingira; former prime minster and DP leader Benedicto Kiwanuka; and Brigadier Shaban Opolot, Amin's former commander—also cheered the demise of the Obote regime. In addition, influential organizations such as the

National Chamber of Commerce and Industry, the Ugandan Teachers' Association, and the Hotel and Allied Workers' Union voiced their support of Amin. Similarly, the country's African and Asian traders applauded the coup because it ended Obote's socialist policies.[13]

To further enhance his popularity, Amin authorized Radio Uganda to broadcast the so-called Eighteen Points, which outlined his predecessor's political and economic shortcomings.[14] He also released many detainees and authorized the return of Kabaka Mutesa's body to Uganda for burial. Additionally, Amin, who maintained that his government was a "caretaker administration," promised free elections and an early return to civilian rule.

It soon became evident, however, that Amin had no intention of relinquishing power. Within days of the coup, Amin promulgated the Armed Forces Decree, which enabled him to assume control of the military. He then abolished the posts of army chief of staff and chief of the air staff, which Obote had created to counterbalance Amin's influence. Amin also consolidated his position in the military by eliminating rivals and promoting supporters to senior positions. Lastly, he restructured the armed forces to ensure their loyalty and reduce chances of a coup. As part of this program, Amin purged Acholi and Langi military personnel, two groups that had prospered under the Obote regime and occupied many influential security posts, and he enlisted an increasing number of foreigners (Sudanese, Nubians, and Zairians) to serve in the rapidly expanding armed forces. (The army increased from about 9,000 in 1971 to approximately 21,000 by 1977.[15])

Amin used Uganda's security establishment to unleash a reign of terror against real and imagined opponents. In early 1971, for example, the Ugandan government enacted a decree that allowed the military to detain anyone, including cabinet ministers, on suspicion of sedition. More significant, Amin created the Public Safety Unit and the State Research Bureau, both of which reported directly to the president's office. Along with the Military Police, these two organizations wreaked havoc on Uganda. By the end of Amin's first year in office, these security forces had killed approximately 10,000 Ugandans. Over the next few years, tens of thousands of Ugandans fell prey to Amin's henchmen, sought sanctuary in neighboring countries, or went into hiding in Uganda.[16]

To deflect public criticism and to enhance his domestic support, Amin adopted a controversial but highly popular program to Africanize the economy. On 9 August 1972, he ordered the expulsion of Uganda's more than 70,000 Asians within three months.[17] The Ugandan government allowed the Asians to repatriate only limited funds and prevented them from receiving any compensation for abandoned property or assets, which included 5,655 firms, factories, ranches, and agricultural estates and about $400 million of personal goods.

The expulsion of the Asian community destroyed Uganda's economy. Most Asian businesses reallocated to African owners quickly collapsed as a result of capital shortages, low inventories, incompetence, and restrictive import policies. Moreover, the economy suffered from shortages of basic commodities such as sugar, soap, bread, milk, and salt. There was a downturn in the production of items such as cement, steel, corrugated iron roofing, blankets, matches, asbestos, cement pipes and sheets, and gunny bags and sacks. By the end of Amin's regime, copper mining and smelting and superphosphate fertilizer production had also ceased.[18]

Apart from its economic implications, the expulsion of Asians crippled Uganda's administrative infrastructure. The departure of hundreds of Asian civil servants caused government services to deteriorate throughout the country. The absence of Asian technical and professional personnel in hospitals, schools, and service industries further inhibited the Ugandan government's ability to care for its citizens.

Shortly after Amin seized power, Uganda's foreign policy entered a period of transition.[19] In little more than a year, Kampala's traditional British and Israeli allies had been supplanted by Libya and, to a lesser extent, the former Soviet Union. This change facilitated the buildup of Uganda's security forces, which plunged the country into chaos.

At the time of the Amin coup Britain dominated nearly every aspect of Uganda's political, economic, military, and social life. However, after the expulsion of the Asians, British-Ugandan relations quickly deteriorated. To express its disapproval of Amin's decision to expel the Asians, Britain cancelled its aid program in Uganda. Amin responded by expelling the British high commissioner and announcing that British citizens had to leave Uganda or remain in the country on local terms. Over the next several months, Amin nationalized thirty-six British firms (January 1973), seized British investments in Uganda (May 1973), and encouraged Ugandans to contribute fruit and vegetables to a Save Britain Fund to help the British government solve its economic problems (December 1973). In June 1975, relations between the two countries suffered another setback when a Ugandan court sentenced British lecturer Denis Hills to death for calling a Amin "a village tyrant."[20] Finally, on 28 July 1976, London broke diplomatic relations with Uganda. This action occurred because of Kampala's failure to explain the death of Dora Bloch, a British subject who had been killed by Amin's soldiers in retaliation for the Israeli rescue of hostages captured by Palestinian terrorists and held at the Entebbe International Airport.

Like Britain, Israel had been one of the staunchest advocates of Ugandan independence. During the 1962–1972 period, more than 1,000 Ugandans studied in Israel. Additionally, Tel Aviv supported Ugandan development efforts by providing advisers to work in several fields, including defense, joint economic enterprises, health, and education.

Initially, Amin, who had received his paratrooper training in Israel, favored close relations with Tel Aviv. In July 1971, he visited Israel and received a $1 million arms sale agreement. Amin also backed Israeli objectives in Sudan by de-nouncing the Arab-dominated Sudanese government for oppressing the black African population in the south and by allowing Tel Aviv to send military assis-tance to southern Sudan's Anya Nya rebels via northern Uganda.

Relations between Kampala and Tel Aviv changed after Amin visited Libya on 13 February 1972 and signed a joint communiqué that pledged both countries to fight against Zionism and imperialism. At the end of February 1972, a Libyan del-egation traveled to Uganda and promised to furnish Amin with military and eco-nomic aid. The day after the Libyans left Kampala, Amin accused Tel Aviv of con-ducting subversive activities in Uganda. On 27 March 1972, he ordered all Israelis to leave the country; three days later, Amin broke diplomatic relations with Tel Aviv.[21]

The expulsion of the Israelis improved Ugandan-Libyan relations. Tripoli helped to resolve the differences between Kampala and Khartoum by arranging an agreement whereby Amin stopped the transshipment of Israeli supplies to Anya Nya rebels. In exchange, the Sudanese government closed an Obote guerrilla training camp in southern Sudan. Libya also increased its economic activity with Uganda. In June 1972, Tripoli agreed to purchase 500 tons of tea and 35 tons of coffee; five months later, the Libyan Arab Uganda Bank for Foreign Trade and Development opened in Kampala with a share capital of about US$8.2 million. Addi-tionally, private business concerns in both countries established the first Arab-Ugandan company to manufacture timber and furniture.

Militarily, Libya offered to train Ugandan Air Force pilots. More important, af-ter Milton Obote's Tanzania-based 1,000-man People's Army invaded Uganda in September 1972, Tripoli deployed five aircraft and 399 technicians to Uganda and promised military assistance to the Amin regime. Despite a cooling of relations in 1974, the Ugandan-Libyan military relationship continued to prosper. In 1975, for example, Tripoli provided 300 military trucks and a squadron of MiG fighter aircraft to Uganda.[22] By the late 1970s, Libya was one of Uganda's most important allies.

Initially, the former Soviet Union showed little interest in the Amin regime, largely because Moscow believed that the coup represented Western interests and posed a threat to Soviet activities in the region. However, after Amin expelled the Israelis, Ugandan-Soviet relations improved. On 9 April 1972, the two countries signed a cultural agreement. Then, on 30 July 1972, a Ugandan military delega-tion visited Moscow and obtained a Soviet promise for security assistance. By November 1973, Moscow had delivered an impressive array of military hardware to Uganda, including MiG-17F fighter aircraft, T-34/T-54 tanks, armored per-sonnel carriers, small arms, and ammunition. Eventually, more than 100 Soviet military advisers taught the Ugandans how to use this equipment. Hundreds of Ugandans also received military training in the Soviet Union. Throughout 1974 and 1975, more than US$500 million worth of Soviet arms poured into Uganda.[23]

Relations between Kampala and Moscow deteriorated after Amin expelled Soviet ambassador Andrey Zakharov because of the Soviet Union's intervention in Angola. Although some Soviet military advisers remained in Uganda, Moscow seemed determined to reduce its commitment to Uganda. However, after the Israeli raid on Entebbe, Amin persuaded the former Soviet Union to resume arms deliveries to thwart Israeli aggression in East Africa.[24]

Although Libyan and Soviet military assistance had enabled Uganda to become one of Africa's major military powers, Amin's hold on power had become increasingly tenuous. Indeed, by late 1978, the Ugandan dictator had laid the groundwork for his demise by eliminating many moderate political and military leaders. His actions intensified rivalries within the armed forces, which destroyed the alliance between foreign elements and Kakwa factions from northwest Uganda that had remained loyal to Amin. Competition between these two groups caused several violent disputes and mutinies within various military units.

To defuse these tensions and to shore up his crumbling government, Amin ordered the rebellious Suicide Battalion from Masaka and the Simba Battalion from Mbarara to annex an 1,800–square kilometer strip of Tanzanian territory north of Kagera River known as the Kagera Salient.[25] Tanzanian president Julius Nyerere responded to this invasion by deploying Tanzania People's Defence Force (TPDF) units to the Kagera Salient. Within two months, the Tanzanians had succeeded in expelling the Ugandans from Kagera.

On 14 November 1978, Nyerere then authorized the TPDF to invade Uganda to oust Amin.[26] As it advanced into southern Uganda, the TPDF recruited approximately 1,000 pro-Obote Ugandans who called themselves the Uganda National Liberation Army (UNLA). The TPDF-UNLA invasion force, which numbered about 45,000, quickly captured the southern town of Masaka near Lake Victoria on 24 February 1979 and then overran Mbarara the following day. Preparations then were drawn up for the final assault on Kampala.

By mid-March 1979, approximately 2,000 Libyan troops and several hundred Palestine Liberation Organization (PLO) personnel had come to Amin's aid. Despite this intervention, however, the joint Tanzanian-Ugandan force continued its assault on Amin's soldiers. Eventually, many Ugandan garrisons deserted or mutinied when they realized that Amin was going to lose the war. Finally, on 10 April 1979, Kampala fell. Amin went into exile in Tripoli, Libya, and about 8,000 of his soldiers retreated into Sudan and Zaire. The following day, Tanzania recognized the interim government of Yusuf Lule, former Makerere University vice-chancellor and chairman of the UNLA's political arm.[27] Approximately 40,000 TPDF troops remained in Uganda to help maintain law and order and to facilitate the emergence of a unified Ugandan government.

Tanzanian soldiers capture one of Amin's henchmen. The Ugandan-Tanzanian war marked a turning point in Uganda's history insofar as it ended the brutal dictatorship of Idi Amin. (Photo by Thomas P. Ofcansky.)

The Sixty-Eight-Day Lule Government

Upon assuming the presidency, Lule promised "to bring back to the people of Uganda the good life they once knew."[28] He also devised a series of policies to restore stability and to rehabilitate Uganda. Lule failed to achieve these goals largely because of his pro-Buganda orientation. Shortly after taking office, he urged his fellow Baganda to unite behind him. Additionally, he appointed conservative Baganda politicians to key ministries, thereby alienating many potential supporters. Some Baganda who served in key ministries included Robert Serumaga (Commerce), Andrew Kayiira (Internal Affairs), and Sam Ssebagereka (Finance).

Because of his political ineptitude, Lule remained in office for only sixty-eight days. His downfall resulted from his inability to establish a working relationship between his government and the Uganda National Liberation Front (UNLF), the UNLA's political wing. In particular, Lule repeatedly clashed with the National Consultative Council (NCC), an organization formed by the UNLA at the March 1979 Moshi conference to act as a parliamentary body.[29]

The most serious Lule-UNLF confrontation occurred as a result of the president's insistence to appoint and dismiss ministers without the NCC's approval. Most UNLF members believed that Lule had made cabinet changes to promote the interests of Baganda traditionalists and conservative businessmen at the expense of more liberal elements. The gulf between Lule and the UNLA widened after five Baganda elders sent a letter to the president claiming that the NCC was subordinate to Lule. His attempt to introduce a quota system for recruitment into the army further alienated the NCC, which feared that the Baganda, the country's largest ethnic group, eventually would dominate the military. The appearance of pro-Lule vigilantes in Buganda proved to be the president's undoing.

On 19 June 1979, the NCC voted to replace Lule with another prominent Baganda politician, Godfrey Binaisa, who had served as attorney general in Obote's first government. Three days later, the UNLF released a statement that outlined the reasons for Lule's removal. The document accused Lule of humiliating and denigrating the NCC and failing to consult with the NCC before making cabinet changes. Lule had also refused to expand the NCC from 30 to 60–70 members for fear that an expanded council would become too powerful. Fourth, Lule violated the NCC's policy that local communities elect their own chiefs by appointing regional commissioners. Additionally, the UNLF statement charged the president with discrimination for allowing the cabinet to award each member Shs. 40,000 in foreign currency as rehabilitation payment while ignoring the plight of other Ugandans who had suffered under Amin. Other problems concerned Lule's failure to devise a policy for the reallocation of houses and businesses, to bring Amin's henchmen to justice, and to prevent the media from criticizing the NCC. Finally, the UNLF accused the president of using the threat of Obote's return to Uganda to justify his dictatorial policies.[30]

Following Lule's dismissal, approximately 2,000 of his followers demonstrated outside Kampala's Nile Mansions Hotel, which served as NCC headquarters. Police fired on the demonstrators, killing two and wounding fifty. Several days later, another protest occurred after authorities arrested the leaders of a pro-Lule group called the Internal Joint Underground. On 26 June 1979, a general strike closed many businesses in Kampala.[31] Despite this popular support, however, Lule failed to regain power. After a brief stay in Dar es Salaam, where he claimed that he had been held prisoner in Tanzania's State House, Lule and Yoweri Museveni formed the National Resistance Movement (NRM) to unseat Obote's second government. In January 1985, Lule died in exile in London.

Godfrey Binaisa

Like his predecessor, Binaisa was a political neophyte.[32] He failed to develop an adequate support base in the UNLF and to exert any control over the increasingly undisciplined UNLA. Binaisa also clashed with the NCC over the issue of the president's right to make cabinet appointments without consulting the NCC. Additionally, he was exceedingly corrupt, even by Ugandan standards. According to his critics, Binaisa "seemed to retain two objectives, to hold on to power and to enrich himself and his cronies."[33] To add to his woes, Idi Amin, who was in exile in Libya, promised to mount a military offensive to regain political power. With large numbers of Amin's followers in Sudan, Zaire, and Kenya, many Ugandans believed this threat was credible.

Despite the many shortcomings of his administration, Binaisa embarked on an ambitious program to restore Uganda's military and political stability. Binaisa pledged to enhance internal security by converting the UNLA into a modern 15,000-man army. The NCC's Standing Committee on Defence and Security assumed responsibility for implementing this policy. Unfortunately, the committee's members failed to agree upon a strategy to reconstitute the armed forces. A majority wanted to equip the army with Western arms and equipment. However, Yoweri Museveni, then defense minister, believed that Uganda should accept military assistance from any source, including the former Soviet Union and its allies.[34] This disagreement, exacerbated by endless political bickering within the NCC, prevented the formation of a new national military force.

To end the lawlessness that plagued Kampala and other larger towns for almost six months after Amin's overthrow, the NCC created the so-called ten-house (Mayumba Kumi) cell system. Under this scheme, the UNLF government appointed a group of ten houses in every village and town to monitor subversive activities and prevent crime. Cell leaders, known as vigilantes, could arrest anyone suspected of antistate behavior. Inevitable excesses associated with this system, which was supported by Tanzanian soldiers who remained in Uganda after Amin's downfall to help stabilize the country, caused many Ugandans to speak out against what they perceived as petty harassment by local UNLF security officials. Even though the ten-house strategy and the presence of Tanzanian military units in many towns and villages restored order to some parts of the country, instability remained part of the Ugandan scene. In July 1979, for example, a Christian-Muslim conflict in Ankole District resulted in more than 80 deaths; three months later, armed gangs killed more than 100 people in Kampala.[35]

In response to these and numerous other incidents, Binaisa mobilized the government against dissident and lawless elements. In September 1979, he authorized a dusk-to-dawn curfew in Kampala District. Additionally, in October 1979 he renewed the Public Order and Security Act of 1967, which allowed for preventive detention. However, even these harsh steps failed to restore Uganda's internal stability, especially in areas like the Karamajong region, where cattle raiders and criminals preyed on villagers.

Politically, Binaisa claimed to be committed to establishing democracy in Uganda. However, on 14 August 1979 he placed a two-year ban on political parties and announced that all political activity had to occur under the UNLF "umbrella." Binaisa believed that this strategy would avoid "the politics of religion, sectarianism, rivalry and hatred."[36] This action naturally prompted protests from political parties and opposition elements throughout the country.

In addition to these difficulties, Binaisa faced several security-related problems. The presence of thousands of Tanzanian troops and police in Uganda created economic hardships for Kampala, which had to pay Shs. 87 million per month for their upkeep. The Tanzanians also became a political liability for Binaisa, who gained the reputation of being a leader kept in office by a foreign power. The police suffered from factionalism. Some units and individuals supported Binaisa whereas others favored Obote. To make matters worse, the UNLA consisted of four competing ethnic elements. One group was loyal to David Oyite-Ojok (Lango), who backed Obote; another supported Tito Lutwa Okello (Acholi), a longtime Obote disciple who now endorsed the DP; the third followed Yoweri Museveni (Ankole), who sought to revolutionize Ugandan society; and the last promoted William Omaria, whose men opposed the Karamojong. The absence of a pro-Binaisa faction in the UNLA proved to be significant.[37]

In March and April 1980, Binaisa's regime began to unravel. Popular opposition to the plan, which required all candidates to run in the December 1980 election under the UNLF "umbrella" rather than as individual parties, further isolated Binaisa. Deteriorating civil-military relations also undermined the president's position. After a series of reports appeared in the *Citizen* and *Economy* weeklies, the military charged the government with leaking the documents and arrested James Namakajo, Binaisa's press secretary, and Roland Kakoza, editor of the *Economy*. Binaisa responded by dismissing the army chief of staff, David Oyite-Ojok.

On 13 May 1980, the Military Commission, which supposedly supervised the army, seized power and placed Binaisa under house arrest. (This was not the end of Binaisa's political career, however: In 1991, he became honorary chairman of a U.S.-based anti-Museveni group known as the Ugandans for Peace and Democratic Pluralism.) The Military Commission, headed by Paulo Muwanga, then scheduled a multiparty election for 10 December 1980. Four parties contested the election: Obote's UPC; the DP, led by Paul Ssemogerere; Yoweri Museveni's Uganda Patriotic Movement (UPM); and the Conservative Party (CP), headed by the *katikiro* of Buganda, J. S. Mayanja-Nkangi. With Muwanga managing the election, the UPC won 74 of the 126 parliamentary seats, and Obote became Uganda's president for the second time. The new government ignored DP and UPM charges of voting irregularities by much of the population; many international observers believed that Obote's supporters had rigged the election.[38]

Obote's Second Government

Obote's second government initially scored some impressive gains, especially in the economic arena. Public and private transport networks again operated throughout much of Uganda; the railway to Mombasa was back in service; the government started rehabilitating the nation's road, water, and electric systems; tea exports resumed; sugar and cotton crops reappeared; the inflation rate declined from 104 percent in 1980 to 30 percent in 1983; and the gross domestic product increased 3.9 percent in 1981, the first evidence of economic growth since 1977.[39] Moreover, Western financial institutions such as the World Bank and the International Monetary Fund (IMF) reestablished links with Kampala.

Eventually, however, Obote not only failed to resolve Uganda's many political, economic, and social problems but also perpetuated the violence and chaos that characterized the Amin regime. This turmoil emanated from Obote's lack of political legitimacy. Like the government of his ruthless predecessor, Obote's government rested on a narrow ethnic base, with most of its supporters coming from the northern Acholi and Lango districts. Like Amin, Obote used military force, intimidation, and terror to silence his real and imagined opponents. By late 1984, these brutal tactics had resulted in the deaths of more than 300,000 Ugandans.[40] Nevertheless, mounting opposition from three groups eventually resulted in Obote's downfall.

The Baganda had been at odds with Obote ever since his first presidency. Their animosity toward Obote started on 24 February 1966, when he abrogated the 1962 Constitution, which had given numerous special powers to Buganda (to raise tax revenues, pass laws, maintain local courts, and preserve its land tenure system). An interim constitution enabled Obote to remove Kabaka Mutesa from the presidency and assume the office himself. When Buganda objected to these tactics, Obote responded by declaring a state of emergency in the kingdom. The rift caused by these actions never healed, and when Obote became president for the second time, the Baganda wanted him removed as soon as possible.

Southern peoples such as the Basoga, Batoro, Banyoro, and Bakonjo also opposed Obote because they despised the northern-dominated UNLA. As the military's involvement in human rights violations increased, these feelings intensified. Many southerners eventually joined or provided support to rebels who fought against the Obote regime.[41]

Numerous insurgent groups posed the most serious threat to Obote, nearly all of which started military operations against the government to protest the rigged presidential election. With the exception of the NRM, all rebel organizations failed to develop a coherent political program. Instead, these groups fought to advance narrow ethnic interests.

For example, former members of Idi Amin's army organized themselves into the Former Uganda National Army (FUNA), led by Major General Isaac Lumago, and the Uganda National Rescue Front (UNRF), both of which were active in the West Nile District. The latter group—led by onetime UPC secretary Felix Onama and Amin's former finance minister, Brigadier Moses Ali—had approximately 3,000 men under arms. The UNRF operated mainly in Terengo County, a remote area between Arua and Moyo. A smaller UNRF force conducted guerrilla activities along the Sudan border.[42]

The Uganda Freedom Movement/Army (UFM/A), created in 1979 by Andrew Kayiira, who had been Lule's minister of internal affairs, and former DP secretary general Francis Bwengye, enjoyed its greatest strength in Buganda. On 9 February 1981, the UFM/A initiated military operations by assaulting five army barracks, seven police stations, and a prison in the Kampala area. Approximately one year later, about 300 UFM/A insurgents launched a mortar attack on Lubira Barracks in Kampala and, according to a government statement, killed four soldiers while losing 67 guerrillas.[43] The government reacted by arresting or killing hundreds of UFM/A members and supporters. Although it remained active, the UFM/A never again posed a serious military threat to the Obote regime.

The smaller Federal Democratic Movement of Uganda (FEDEMU) also had a large following in Buganda. David Lwanga, who eventually became minister of environmental protection in Yoweri Museveni's regime, led this group. FEDEMU's military wing, the Federal Democratic Army, included remnants of the UFM/A and the UNRF.

The most successful of the guerrilla groups was the National Resistance Movement/Army (NRM/A), which had started as the Popular Resistance Army after merging with a small insurgent group known as the Ugandan Freedom Fighters. The movement, which was based in Buganda's Luwero, Mpigi, Mubende, and Mukono districts, included Baganda, Bairu, Bakiga, and Banyoro troops.[44] Yoweri Museveni—who had received military training in Mozambique with the Front for the Liberation of Mozambique, had gained battlefield experience against Idi Amin's soldiers, and had served in Binaisa's government—commanded the NRA. Museveni's effectiveness as a guerrilla leader depended on his ability to attract financial aid and military assistance from sources as diverse as Libya and the British conglomerate Lonrho. On 6 February 1981, the NRA commenced guerrilla operations against the government, which initially were confined largely to the Luwero triangle, north of Kampala between the main roads north to Hoima and to Gulu. However, to conserve his forces, Museveni avoided pitched battles with the UNLA; instead, he attacked trains, buses, lorries, and convoys of coffee trucks to weaken the government's control over the country's infra-

structure. The NRA financed its operations by robbing banks and accepting do-
nations from ex-government officials and ordinary Ugandans. Many local communities also supplied Museveni's troops with food, shelter, and information
about UNLA movements.

By January 1983, the NRA controlled approximately 4,000 square miles of territory north and northwest of Kampala. Because of its battlefield successes, the
NRA attracted more and more recruits, including a large number of UNLA deserters. In August 1983, the movement had about 6,000 personnel; the following
year, it had grown to approximately 9,000 soldiers. This growth enabled Museveni
to open a second front in western Uganda's Ruwenzori Mountains.

To enhance the NRA's prestige, Museveni claimed that he maintained discipline
and morale among his troops by adopting a nonsectarian policy that allowed
Ugandans from all ethnic groups to join his movement. He also stated that he ensured that NRA personnel respected human rights (the penalty for killing civilians was death) and understood the movement's political goals as outlined in
Museveni's "Ten Points." Moreover, he continually publicized the fact that the
NRA was a "people's army," that its officers were "politicians in uniform," and that
no one in the movement wore badges of rank. Despite this carefully cultivated
reputation, however, the NRA engaged in some questionable activities. According
to one report, its record was no better than the other antigovernment insurgent
groups.[45]

Several factors prevented the Obote regime from defeating these insurgencies.
By June 1981, most TPDF units had withdrawn from Uganda; the few Tanzanian
soldiers who remained in the country devoted their time to train the UNLA
rather than engaging in counterinsurgency operations. As a result, Obote had to
fight an array of rebel groups with an army that was undisciplined, ill-trained,
poorly equipped, and badly commanded. Obote also put himself at a political disadvantage by relying on the northern Acholi and Langi peoples to fill UNLA
ranks. This allowed his enemies to criticize him for using a strategy whereby the
Acholi could hold other ethnic groups to political ransom and commit violence
against them. Obote tried to improve the UNLA's military capabilities by using
foreign military advisers. Apart from the Tanzanians, Soviet and North Korean
personnel conducted a variety of educational programs and a British security
company, Falconstar, trained a paramilitary special force. In April 1982, a 36-man
Commonwealth team arrived in Uganda to train UNLA units, largely at Jinja barracks. These few foreign military advisers failed to make much of an impact on
the 35,000-man UNLA, and it remained an undisciplined force. In mid-1984, the
Commonwealth countries withdrew their personnel because of the UNLA's poor
human rights record. Great Britain decided to keep a 12-man team in Uganda.

As a result, the UNLA conducted its operations with little regard for the rules
of warfare. As a result, terrible human rights abuses occurred against pro-NRA
communities in the Luwero triangle. In January 1983, Obote launched "Opera-

tion Bonanza" in this area, during which UNLA troops destroyed small towns, villages, and farms and killed or displaced hundreds of thousands of civilians. The carnage eventually attracted the world's attention, and several governments and humanitarian organizations condemned the Obote regime. According to Amnesty International, there were reports of at least thirty-six mass grave sites in the Luwero triangle. The Banyarwanda community, much of which had supported Amin, lost 45,000 to 60,000 people. After the war ended in 1986, the International Committee of the Red Cross claimed that at least 300,000 people had died in the Luwero triangle and that officials had failed to account for half to a third of the region's population.[46]

There also were large-scale human rights violations in other areas of the country. In late 1982 and early 1983, groups of local chiefs and UPC youth wingers, supported by Special Forces personnel, unleashed a reign of terror against the pro-DP Banyarwanda and Bahima peoples in southwestern Uganda. As a result of widespread killings, rapes, and maimings and the destruction of about 16,000 homes, approximately 35,000 people resettled into government-protected villages and another 40,000 escaped to Rwanda. Similar attacks occurred against the Banyarwanda in Teso and Lango in eastern and northern Uganda.[47]

Security personnel also assaulted Ugandans who tried to return to their homes in West Nile District from refugee camps in southern Sudan. Some UNLA units even launched operations against refugee camps in southern Sudan, which supposedly housed anti-Obote elements. By mid-1984, these actions had resulted in approximately 120,000 deaths. Additionally, thousands who lived in Karamoja and Kampala perished at the hands of UNLA soldiers or bandits.

On 23 May 1984, UNLA troops massacred 75 to 300 students, teachers, and priests who studied at the Namugongo Anglican Theological College and Seminary after guerrillas had attacked a ground satellite station at nearby Mpoma and supposedly had taken refuge among the seminarians. This incident caused a rupture in relations between Uganda and the United States. Various U.S. officials, including the assistant secretary of state for human rights, Elliott Abrams, and the U.S. ambassador to Uganda, Allen Davis, accused the Obote regime of killing, jailing, or torturing innocent civilians throughout the country and exterminating 100,000 to 200,000 people in the Luwero triangle. Kampala reacted to these charges by expelling Colonel Hugh Baker, a visiting U.S. military attaché, and canceling a $100,000 training program for UNLA officers in the United States. West Germany, Denmark, and Holland supported the U.S. position.

The British government, however, adopted a less critical policy toward Obote. Whitehall rejected the bleak U.S. assessment of the Ugandan situation. In August 1984, the British government launched its own investigation of Obote's human rights record; at the same time, it agreed to keep the twelve-man British military training team in Uganda. After Amnesty International released a report entitled *Uganda: Six Years After Amin* in June 1985, the minister of state for African affairs at the Foreign and Commonwealth Office, Malcolm Rifkind, cautioned the Ugandan high commissioner that if the human rights situation in Uganda did not improve, Britain might terminate its £7 million annual assistance program to Uganda. By the time of this threat, Obote was about to fall prey to a military coup.[48]

In addition to growing international criticism, at least two other factors eroded Obote's legitimacy. Since late 1984, Vice-President and Defense Minister Paulo Muwanga had been plotting against Obote. Apart from objecting to the president's plan to name his cousin, Akena Adoko, chairman of the Public Service Commission, Muwanga wanted to negotiate with the insurgents—unlike Obote, who believed he could achieve a military victory over his opponents.[49]

Although these differences slowly weakened his regime, Obote lost power because of his inability to preserve the fragile Acholi-Langi alliance, especially in the UNLA. Many Acholi believed that Obote favored his fellow Langi in new military appointments and promotions. In August 1984, this conflict escalated when Obote named Smith Opon-Acak, a Langi, to the post of chief of staff. This left the soon-to-retire seventy-one-year-old General Tito Lutwa Okello as the only Acholi in a key military position. As soon as Opon-Acak took office, Acholi personnel accused him of unfairly advancing the careers of his fellow Langi and of deploying only Acholi troops to combat zones.

Soon afterward, disaffected Acholi soldiers started plotting with opposition leaders to overthrow Obote. Troops in the ranks also began disobeying their Langi officers. In June 1985, for example, soldiers assigned to Magamaga Ordnance Depot refused to go into combat against the NRA in western Uganda. A few weeks later, interethnic fighting at Mbuya Barracks, which came about when it became known that Obote had ordered the arrest of several Acholi officers, claimed the lives of at least thirty UNLA personnel. To prevent a future purge or massacre of Acholi military personnel, Brigadier Basilio Okello, an Acholi, mobilized antigovernment UNLA troops at his Gulu headquarters and marched on Kampala to overthrow Obote. Along the way, he defeated pro-Obote Langi forces at Karuma Falls and at Bombo. Finally, on 27 July 1985, Brigadier Basilio Okello and his men entered Kampala, seized Radio Uganda, and announced that Obote's regime had come to an end.[50]

The Okello Regime

Upon assuming office as head of state, General Tito Lutwa Okello, an Acholi, claimed he had participated in the overthrow of Obote to end the fighting, facilitate the peace process, and promote human rights. He also issued a statement informing Ugandans that he had suspended the constitution, dissolved parliament, dismissed all ministers, temporarily closed the country's borders, and outlawed all foreign exchange transactions. Additionally, he promised to create a government of national unity. To achieve this goal, he urged all political parties and insurgent groups to join the regime. In August 1985, FEDEMU, FUNA, UFM, and UNRF accepted this invitation and joined the ruling Military Council. Significantly, the NRM/A refused to participate in the Okello government, largely because five of the nine Military Council members were Acholi. (The Acholi members included Tito Lutwa Okello, chairman and head of state; Basilio Okello, army commander; Justin Okot, secretary of the council; Colonel Fred Oketcho; and Colonel Jack Nyero. The other members included Lieutenant-Colonel Sam Nanyumba, Colonel Zedi Maruru, Colonel Gurd Goodwin Toko, and Lieutenant-Colonel Fred Ocero.) Museveni argued that at least half of the Military Council seats and a corresponding number of cabinet seats should be filled by NRM/A personnel. He also opposed the creation of any other political authority. Okello rejected both demands. To establish continuity with the previous government, he then appointed Paulo Muwanga, who had served as Obote's minister of defense, as prime minister and directed him to form a cabinet that would be subordinate to the Military Council. Muwanga's appointment was short-lived, largely because he bore responsibility for many of the human rights violations committed during the Obote years. When Museveni responded to Muwanga's appointment by initiating military operations against the Okello government, the Military Council named Abraham Waligo as the new prime minister. Okello also sought to strengthen his position by inviting former soldiers of Amin's army to reenter Uganda from refugee camps in southern Sudan and join the government's campaign against the NRA. This tactic backfired insofar as it only enhanced Museveni's growing reputation as a leader who would bring genuine change to Uganda. With his government crumbling, Okello agreed to peace talks with the NRA in Nairobi, Kenya.[51]

Kenyan president Daniel arap Moi chaired the negotiations, which lasted between 26 August and 17 December 1985 and ended with a peace accord between Yoweri Museveni and Tito Lutwa Okello. The agreement's terms included a ceasefire; the creation of a Military Council comprised of Tito Lutwa Okello as chairman and head of state, seven representatives each from the UNLA and NRA, and five representatives each from the smaller groups (i.e., FEDEMU, FUNA, UFM, and UNRF); the creation of an 8,480-man national army that included 3,700 from the UNLA, 3,580 from the NRA, and 1,200 from the other insurgent groups; the demilitarization of Kampala; the establishment of a multinational force, made up of troops from Kenya and Tanzania, to supervise the implementation of the peace agreement; a review of all government decrees and appointments made since July 1985; the convening of a national conference to prepare for an interim government and elections; and the abolition of the NRA's administration in southwestern Uganda.[52] Since the peace accord was never implemented, instability continued to plague Uganda.

On 31 December 1985, just two weeks after the two sides had agreed to stop fighting, Tito Lutwa Okello accused the NRA of sabotaging the agreement by launching a military offensive against the UNLA at Kabasanda on the Kampala-Masaka road, obstructing food relief supplies en route to the garrison at Mbarara, impeding the travel of the multinational force, and failing to nominate its representatives to the Military Council. Museveni responded by claiming that the UNLA had killed 240 people and had shown favoritism in military promotions since the two sides had signed the peace accord.

Hostilities continued unabated between the two groups. By early January 1986, Museveni claimed that the NRA controlled a region that produced 60 percent of Uganda's coffee. He also proceeded with plans to seize Kampala and overthrow the Military Council. At Jinja, 1,000 UNLA soldiers surrendered to the NRA. As Museveni's troops approached the capital, thousands of noncombatants fled the city. The UNLA also retreated, leaving devastation in its wake. Finally, on 26 January 1986, the NRA succeeded in seizing control of Kampala and established itself as the government of the country.

Museveni's Government

Yoweri Museveni's presidency has been an important turning point in Uganda's postindependence history. Under his leadership, the country gradually has moved away from the brutal violence of past regimes, although his government still has an objectionable human rights record. There has also been noticeable progress in reducing ethnic tensions in some parts of Uganda. In the economic sphere, Museveni has succeeded in acquiring impressive amounts of foreign economic aid from several Western nations and international financial institutions such as the World Bank and the International Monetary Fund. These accomplishments have helped to establish Museveni as one of Africa's foremost leaders. As chairman of the Organization of African Unity (OAU) from 1990 until 1991, he earned a reputation as a trusted and effective international statesman. Even after eight years in office, he enjoys considerable popularity among Ugandans from all walks of life.

Despite his many accomplishments, Museveni must contend with the realities of a desperately poor country with an abundance of political, economic, and social problems. To resolve these difficulties, he relies on a simple but seemingly effective ideology, which he summarizes by saying, "We take from every system what is best for us and we reject what is bad for us. We do not judge the economic programmes of other nations because we believe that each nation knows how best to address the needs of its people. The NRM is neither pro-West nor pro-East: it is pro-Uganda."[53] In practical terms, this ideology enables Museveni to maintain close ties with nations as disparate as the United States and Libya; reject Western-

Liberation Day, Kampala, 26 January 1986. The enthusiasm of that day continues to live on throughout the country as all Ugandans are determined to improve their lives. (Photo by Thomas P. Ofcansky.)

style multipartyism as destabilizing; create what probably is Uganda's most effective postindependence grass-roots political system but threaten to kill anyone who attended a rally sponsored by a faction of the opposition DP; and take pride in the fact that Uganda has one of the freest presses in Africa but continue to persecute journalists for publishing articles critical of the government.[54]

In terms of his leadership qualities, Museveni is a man of great energy and vision. However, he has the reputation of interfering in the day-to-day work of government agencies, which often demoralizes the staff. In the foreign policy arena, Museveni has a remarkable ability to maintain good relations with the West and international financial institutions, despite his opposition to Western-style democracy, rampant corruption throughout his government, and a human rights record that leaves much to be desired.

On 29 January 1986, Museveni displayed a confident optimism in his leadership traits when he was sworn in as the new president of Uganda. In his inaugural speech, he promised to establish democracy, promote unity, and end human rights violations. He also pledged to abolish his "interim administration" within four years in favor of an elected government under a new constitution. These and other goals were included in the NRM's Ten-Point Program. The goals of this program included establishing popular democracy, restoring stability, encourag-

ing national unity, maintaining national independence and nonalignment, rebuilding the economy, creating a mixed economy, restoring and rehabilitating social services, ending corruption and misuse of power, resettling displaced people,
and promoting regional cooperation and human rights. Achieving these objectives has proven to be an elusive undertaking. Although the Museveni regime has
made some progress toward ending the military, political, and economic turbu-
lence of the past several decades, Uganda remains a deeply troubled society.

Since seizing power, Museveni has used a multifaceted strategy to create what
he claims is a democratic government that relies on consensus rather than force.
On 1 February 1986, for example, he changed the Military Council into a National Resistance Council (NRC), which contained twenty-one military and civilian
members. Museveni served as NRC chairman. He claimed that this government,
which he admitted was not elected, reflected his commitment to democracy. To
ensure popular support, Museveni formed a coalition government, which included FEDEMU, DP, UPC, UNRF, and UFM. All of these groups eventually withdrew from the coalition, citing the government's complicity in human rights violations, official corruption, continuing instability in northern and eastern
Uganda, the creation of tribal animosities, and communist and Libyan infiltration
of Uganda. (The DP later rejoined the government.) It is interesting to note that
in one of his first official acts, Museveni announced that he had postponed elections for four years, thereby cancelling the elections Obote had scheduled for
the end of 1985. He also banned political parties from holding meetings without
government approval.

On 19 February 1986, the Ugandan government issued a statement outlining
the "Structure of the Resistance Councils and Committees." According to
Museveni, this concept organized every village (and in the case of Kampala and
larger towns and municipalities, every ward of no less than ten and no more than
twenty families) into a Resistance Council (RC I). By late 1987, there were about
40,000 RCs at the village level. From this level, RCs are arranged in an ascending
hierarchy to cover the parish (RC II), subcounty (RC III), county (RC IV), and
district (RC V) levels. Each RC has a nine-person elected executive committee
that has the political and judicial authority to manage local affairs. The national
legislature, known as the National Resistance Council, is composed of 278 members, 212 of whom are elected by RCs at the county and district levels. Representation through the RCs culminates in the National Executive Committee (NEC),
which includes some people who had been elected at the district level.[55]

According to the NRM, the RC system is the vehicle by which all Ugandans are
brought into the national political system. Additionally, the RCs, which are involved in development projects, disease control, famine relief, and preventing desertification and soil erosion, also give people an opportunity to influence the
course of their daily lives. RCs also guard against official corruption and misuse of

power. Most important, the NRM maintains that the RC system facilitates the growth of representative democracy at the local level.

Although they afforded countless Ugandans the opportunity to determine their future, the RCs failed to create a nationwide popular, efficient local-level government. Anti-Museveni insurgents often killed RC members to discredit and weaken his regime. Corruption and incompetence plagued many village RCs. Many government bureaucrats opposed the RC system because they feared it would erode their power.[56]

Despite its many shortcomings, the RC system enabled the Ugandan government to hold a relatively free and fair election for an expanded NRC. Between 14 February and 4 March 1989, all Ugandans, with the exception of those who lived in the north and in part of northeastern Teso, participated in the country's second election since independence. According to government guidelines, candidates had to reside in the constituency where they were standing for election. No deposit was required and there was no prior registration of voters. The NRM prohibited party and individual campaigning, however.

Although it increased the government's popularity and cost several cabinet ministers their NRC seats, the election failed to alter the power structure. With few exceptions, NRC members belonged to government, business, and professional elites that traditionally had dominated Ugandan politics. Moreover, the timing of the election, which occurred before the Constitutional Commission, created in November 1988 to draft a new constitution, had started its deliberations, caused many Ugandans to question the NRM's long-term political strategy. On 31 December 1992, the Constitutional Commission finally submitted a draft "no-party" constitution to President Museveni, which recommended suspending party activity for at least seven more years after which a referendum would be held on the matter. The public's skepticism proved to be warranted. On 11 October 1989, the NRC passed a bill that extended the interim period of the NRM government for five years to January 1995.[57]

Even before the NRC gave the NRM a new lease on life, Museveni had announced his view that a multiparty political system was not necessary to ensure democracy in Uganda. Government functionaries also defended the concept of a "no-party" democracy by pointing to irregularities associated with the 1961, 1962, and 1980 general elections and the relative absence of abnormalities in the NRM's 1986, 1989, and 1992 elections. In early 1992, Museveni resisted international pressure to lift the ban on party politics. (Parties such as the DP and UPC continued to exist but were not allowed to have their members run as party candidates.) On 18 March 1992, the president tried to deflect this criticism by promising that there would be a general election in 1994–1995, during which every political leader would be elected by a majority vote. Anti-Museveni elements, however, claimed they would oppose such an election if other political parties were not allowed to sponsor candidates.[58] According to some observers, Museveni hoped to guarantee his political survival by devising a scheme whereby the NRM would be abolished in favor of a new broad-based coalition.

However, by 1993 it had become clear that Museveni wanted to continue a "no-party" democracy for at least another fifteen years. But multiparty advocates wanted a commitment from Museveni that the government would sanction multiparty politics within four years. During the campaigns leading up to the 288-seat Constituent Assembly elections on 28 March 1994, many candidates argued that the "no-party or multiparty issue" eventually should be put to a referendum.[59] In mid-December 1993, this controversy took a surprising turn when a High Court judge ruled that the ban on political parties be tested in a Constitutional Court. Museveni responded by promising to consider asking parliament to amend the law if Uganda's courts ruled that his ban on political parties was unconstitutional.

Another important election issue concerned the status of Uganda's former kingdoms (Buganda, Bunyoro, Ankole, and Toro), all of which had been abolished by the 1967 constitution. From 1986 until early 1993, NRM leaders repeatedly indicated that "they did not fight in order to restore monarchs."[60] However, the Ugandan government feared that the former kingdoms problem could become an issue in the Constituent Assembly election and the 1995 general election. Therefore, President Museveni strongly supported the passage of the Constitution (Amendment) Statute 1993 and the Traditional Rulers (Restoration of Assets) Statute 1993, which authorized the restoration of the traditional rulers in each of the kingdoms. Since then, Museveni repeatedly has warned the restored traditional rulers not to use their positions to seek or assume national powers.

As expected, Museveni's supporters won more than two-thirds of the seats (145 out of 214) in the Constituent Assembly election. (Candidates stood as individuals rather than party candidates.) The remaining 74 members were nominated from women's groups, the military, political parties, and by the president. About 7.2 million Ugandans voted in the first secret-ballot elections in fourteen years. According to Solomon Bosa, head of the Uganda National Election Monitoring Body, the election was marred by faulty voter registration and too much government involvement. However, the leader of a 109-member United Nations (UN)–sponsored observer team, Yilma Tadesse, claimed the elections went smoothly. The Constituent Assembly will create a 200-member national body that will debate a draft constitution, decide whether the country will have a multiparty political system, prepare the way for the 1995 general election, and determine how power will be divided among Uganda's four former kingdoms (Buganda, Bunyoro, Ankole, and Toro).

Apart from trying to resolve the multiparty controversy, Museveni has struggled to restore stability to Uganda. This has been a daunting task. Indeed, despite

Luwero triangle: The cost of war. It is unlikely that Ugandans will ever forget the devastation and brutality that plagued their country during the 1970s and 1980s. (Photo courtesy of U.S. Committee for Refugees.)

repeated government claims that the NRA had defeated all rebel groups, insurgent activity continued, especially in the northern, eastern, and western regions. None of these insurgent organizations presented a coherent alternative to the Museveni regime; many were little more than bandit groups. Throughout the 1986–1993 period, Museveni employed a dual strategy of offering peace agreements or unconditional amnesties and intensifying military operations. In April 1988, for example, 3,000 former Uganda People's Army (UPA) fighters and members of several other small rebel groups accepted an amnesty by surrendering and declaring their support for Museveni's regime. In June 1988, the president concluded a peace agreement with Uganda People's Democratic Army (UPDA) commander Lieutenant Colonel John Angelo Okello (no relation to former president Tito Lutwa Okello). Although the NRA subsequently integrated many UPA and UPDA personnel into its ranks, thousands of others rejected the peace accord and continued to fight against the NRA. Eventually, many of these individuals joined other insurgent groups, turned to banditry, or returned home. As a result, by 1989 the UPDA was finished as a fighting force.

In 1988, the government promised to pardon rebels who lacked criminal records if they surrendered; those who refused would be tried as "bandits" before special courts designated to deal with insurgents. Then, in February 1989,

Museveni declared a three-month moratorium on military operations against rebels near Gulu. However, only a few rebels relinquished their arms. Once the moratorium expired, the NRA launched assaults on rebel bases and in mid-1989 implemented a "scorched-earth" policy in the region. Troops moved several thousand civilians to government-run camps and then burned houses, crops, and granaries in these depopulated areas. In February 1990, the NRA tried to isolate rebel forces by rounding up some 200,000 civilians and placing them in guarded camps in eastern and western Uganda. This tactic enabled the NRA to establish control over some regions. However, it also eroded the government's domestic and international support, largely because of the high number of deaths resulting from inadequate food, water, shelter, and medical care in the camps. Throughout 1991 and early 1992, counterinsurgency operations continued in northern, eastern, and western Uganda.[61]

By mid-1992, several rebel groups, including the Uganda People's Army (UPA), the United Democratic Christian Movement (UDCM, earlier known as the Holy Spirit Movement [HSM]), and the National Army for the Liberation of Uganda (NALU), remained active against the government. Although none of these organizations posed a serious threat to the Museveni regime, their activities undercut the credibility of the president, who repeatedly claimed that the NRA had defeated them, caused widespread suffering among the civilian population, and delayed or prevented numerous social and economic development projects.

Insurgent activities dissipated considerably in late 1993 and early 1994, as rebel groups lost their leaders, surrendered to the government, or simply stopped fighting. On 18 August 1993, for example, Kenyan police found the bullet-ridden body of NALU leader Amon Bazira along the Nairobi-Nakuru highway. The Kenyan authorities never made any arrests in connection with this crime. On 10 January 1994, the Ugandan government opened peace talks with the Lord's Resistance Army (LRA), which originally had been known as the HSM and then as the UDCM. The following day LRA leader Joseph Kony agreed to surrender to the government. However, as of mid-1994 he and a small group of his followers remained active against the government in remote parts of northern Uganda.[62] After the Sudanese government increased aid to Kony in late 1994 to discourage Uganda's support of the SPLA, the rebels increased their operations in the north. However, the LRA still posed no threat to the survival of the Museveni regime.

Although these developments augured well for Museveni and the NRM, they cannot be interpreted as an end to the country's instability problems. Uganda is located in a region that suffers from endemic violence, especially in remote areas where government authority is weak. Moreover, Museveni, although popular with most Ugandans, has yet to build a national political consensus or to establish a broad-based government that has a good chance of surviving his tenure. Given these difficulties, it is likely that Uganda will continue to suffer from internal instability for the foreseeable future.

Another priority issue for Museveni has been his program to end human rights violations. To be sure, there are far fewer crimes against civilians than during the Amin and Obote II regimes. However, when judged on its own merits, the Museveni regime has been a gross violator of human rights, especially in the early years of his regime. During the 1986–1991 period, for example, the NRA expanded tenfold because of the absorption of personnel who had been in the armies of previous regimes or in one of various rebel groups. Many of these individuals were extremely undisciplined and were responsible for committing widespread human rights violations such as murder, torture, and rape throughout the country. Similarly, the police, which also grew during the late 1980s and early 1990s, earned the wrath of human rights advocates for similar activities. Promises by the government to punish those found guilty of committing atrocities have never resulted in the incarceration of senior military or government officials who bear ultimate responsibility for the actions of those under their command. Instead, the authorities have usually jailed or executed low-ranking soldiers. On the positive side, the NRM created several agencies that worked to improve Uganda's human rights record.

In May 1986, for example, the NRM created a Commission of Inquiry into Human Rights Violations to investigate the activities of all governments since independence until the day before the NRA took control of Kampala. The commission examined judicial and other records regarding arbitrary arrest and detention and torture and executions. In December 1986, the commission started selecting witnesses who would testify in public session. One of the most controversial witnesses, a former NRA political instructor, testified that political opponents were considered traitors.

A lack of resources hampered the commission's performance. Financial and transportation problems initially confined its activities to Kampala. Eventually, however, these difficulties temporarily brought public hearings to an end. Although a February 1988 Ford Foundation grant enabled the public hearings to resume, the commission's final report was unavailable in late 1994.

In 1987, the president also established the post of inspector general of government (IGG) to investigate individual complaints about human rights abuses committed since the NRM came to power. The IGG answered only to the president and had the authority, with presidential approval, to seize documents, subpoena witnesses, and question civil servants. Any government official who refused to cooperate with the IGG faced a three-year prison term or a fine. Budgetary problems and staff shortages reduced the IGG's effectiveness, causing an increasing number of Ugandans to complain that investigations were too slow and produced no results.

Several nongovernmental human rights organizations also worked to improve conditions in Uganda. The Uganda Human Rights Activists (UHRA), for example, has monitored developments in Uganda since the early 1980s. Initially, the

UHRA's relations with the Museveni regime were tense, largely because the authorities had arrested Lance Muwanga, UHRA secretary-general, for comparing the NRM's human rights record to that of the Amin government. Since Muwanga's arrest, the UHRA's relations with the government have improved. Currently, the UHRA publishes a quarterly magazine, *The Activist,* which reports on the human rights situation in Uganda. In late 1990, UHRA issued a report that generally approved of Museveni's human rights record.

The Uganda Law Society has been one of the most vocal advocates for protection of human rights in Uganda. Two hundred of the country's 800 lawyers belonged to the Uganda Law Society in 1991. Apart from speaking out against human rights violations in northern and eastern Uganda, the Uganda Law Society has called for an independent judiciary, an end to illegal arrests and detentions, legal reform, and constitutionalism. A lack of funds and resources has hampered Uganda Law Society activities.

The Uganda Association of Women Lawyers, also known as FIDA-U, has worked to make rural people aware of their legal rights, promote family stability through legal advice and counseling, ensure equal protection under the law for women and children, and promote the welfare of all Ugandans by emphasizing laws that stimulate economic development. In March 1988, FIDA-U opened a legal clinic to help indigent Ugandans, especially women and children. By August 1990, FIDA-U had handled more than 1,000 cases that dealt with property rights, inheritance, affiliation, marital, business, and land matters.[63]

To counter accusations of human rights abuse, particularly in northern and eastern Uganda, the Museveni regime punished members of the NRA convicted of assault or robbery against civilians. Several soldiers were executed for murder or rape. Military officers even carried out some of these executions in the areas where the crimes were committed, inviting local residents to witness the executions. Despite protests by several international organizations, these executions continued in 1990. Uganda's attorney general, George Kanyeihamba, justified the practice, insisting that strict discipline was necessary to maintain order in the military.

Despite these harsh measures, human rights violations continued in parts of northern, eastern, and western Uganda in the late 1980s and early 1990s. In October 1987, for example, witnesses reported that soldiers killed 600 people in Tororo District during an NRA counterinsurgency operation. People in the southwest claimed that the security services killed a number of schoolchildren in antigovernment protests and that as many as 200 villagers were shot for refusing to attend a political rally. Murders of people suspected of being rebel sympathizers were also reported.

In early 1989, Dr. H. Benjamin Obonyo, secretary-general of the antigovernment Uganda People's Democratic Movement (UPDM), corroborated some of the evidence acquired by Amnesty International and other human rights organizations. He also charged that the NRA had "burned or buried civilians alive" in regions of the north and east. Obonyo accused three NRA commanders of atrocities against civilians in several districts. Then, in March 1989, the UHRA claimed that the NRA had killed more than 100 civilians during a counterinsurgency campaign in and around Gulu.

Throughout 1990, according to Amnesty International, the NRA killed unarmed civilians in the districts of Gulu, Tororo, Kumi, and Soroti. Amnesty International suspects that some of these incidents were extrajudicial executions. Despite several government inquiries, Amnesty International claimed that no NRA personnel alleged to have committed these human rights violations were ever charged or brought to trial. Moreover, more than 1,300 people remained in detention without charge or trial at the end of the year. On 4 December 1991, Amnesty International released a twenty-one-page report that outlined the abuses committed by the NRA during a nine-month counterinsurgency campaign in northern Uganda.[64] This was the largest report Amnesty International ever published on the Ugandan army. Kampala denied the accusations contained in this document.

In September 1992, Amnesty International published another report, which urged the Ugandan government to end extrajudicial executions, torture, and unlawful detentions and to punish soldiers and government officials guilty of human rights violations. The Ugandan government responded to Amnesty International by issuing a statement that described the report as "politically biased and out of date."[65]

After the appearance of these two reports, growing international and domestic pressure persuaded President Museveni to disavow the use of brutal tactics and to improve the country's human rights record. Conditions gradually improved, and by late 1993 Museveni had released almost all political prisoners and had convinced many Western nations that he was a strong human rights advocate. However, human rights abuses continued to occur in many parts of Uganda. Such incidents were perpetrated by NRA personnel, many of whom acted in contravention to standing orders that sought to protect the rights of innocent civilians. Also, several individuals who worked in the Ministry of Justice, which was known in some Ugandan circles as the Ministry of Injustice, continued to commit outrages against real and imagined opponents of the Museveni regime. Moreover, the Museveni regime continued to maintain restrictions on organized political party activity, harass journalists by charging them with sedition, and torture and beat some prisoners.

There is no question that there has been a significant improvement in Uganda's human rights record since 1986. Indeed, by the mid-1990s, the government's "forgive and forget" attitude toward former guerrillas, dissidents, and members of previous regimes enabled many Ugandans to return home after years of exile without fear of official retribution. There also had been a reduction in reports of government abuse of suspected rebels and criminals. As a result, by late 1994, Uganda had one of the better human rights records in East Africa, especially when compared to neighbors such as Sudan, Zaire, and Rwanda during the final months of the Habyarimana regime. However, given the ethnically fragmented nature of Ugandan society and Kampala's inability to restore stability to some parts of the country, preserving human rights will remain an important issue for the foreseeable future.

4

SOCIETY AND CULTURE

Historically, diversity has characterized Ugandan society and culture. Over the centuries, peoples of varying linguistic, religious, and ethnic backgrounds from Africa, Europe, and India settled in Uganda. Although a source of strength in some other countries, this diversity helped to destabilize Uganda as different ethnic groups competed for political, economic, or military advantage. The resulting social and cultural divisions allowed indigenous rulers and British imperialists to preserve or establish political power by playing one group off against another. Ugandan society and culture has also suffered from at least two large-scale medical catastrophes: A 1901–1905 sleeping sickness epidemic claimed hundreds of thousands of lives, and acquired immune deficiency syndrome (AIDS) of the late twentieth century threatens the well-being of the entire population.

The current government, which seized power in 1986, is committed to creating a single national identity without extinguishing the country's social and cultural diversity. To achieve this goal, Uganda's leadership will have to reverse centuries of ethnic divisions and to establish a political system that includes adequate representation for all ethnic groups. However, given the country's proclivity toward social and cultural divisiveness and its lack of resources to create a genuine national government, it is unlikely that Uganda will achieve these goals anytime soon.

This chapter will review the main elements that have affected the development of Uganda's social and cultural heritage. The first section will examine recent demographic trends. Other sections will examine language and ethnicity, religion, education, health, women, and popular culture.

Population and Demography

Determining the number of people in Uganda has always been an imprecise exercise. Even in the best of times, a lack of government resources prevented officials from conducting an accurate census. That many Ugandans lived in remote rural regions only exacerbated the problems that confronted census takers. Over the years, however, there have been some methodological improvements. During the 1911 and 1921 censuses, officials counted the African population on a county (*saza*) basis. Beginning in 1931, the authorities used the village (*muruka*) rather than the *saza* as the basis of a census. The 1948 census witnessed the limited use of more detailed questionnaires that gave a more accurate picture of the social and economic condition of Uganda's population. Using information gathered during the 1948 census, demographers made a population density map that the authorities used in the 1959 census. After the 1969 census, the coding and processing of the collected data were done in Uganda for the first time. The 1980 census was controversial because field results from many regions were lost or stolen. As a result, any national-level analysis of the population was impossible.

In January 1991, the Ugandan government conducted its latest census, which estimated the country's population to be 16,582,600 people (1,882,600 in urban centers and 14,700,000 in rural regions). International estimates ranged as high as 19 million. (Table 4.1 summarizes Ugandan population growth since 1911.)

Census officials contended that the 1991 census figures reflected a "fairly accurate count." Those who maintained that Uganda's population is close to 19 million argue that instability in the north and northeastern parts of the country prevented a correct counting of the population.[1] The former estimate reflected an annual 0.2 percent per annum drop in the population growth rate over the past eleven years. However, in real terms, the population increased by almost 4 million during this period.

According to the Ugandan government, the 1991 census represented the third downward trend in the country's population growth since the first nationwide census in 1911. In 1948, the population growth rate dropped to 2.0 percent from 2.1 percent in 1931 and to 2.8 percent in 1980 from 3.7 percent in 1969.[2] Some demographers claim that the 1948 and 1980 indications of decline resulted from statistical errors. However, other demographers explain the 1980 and 1991 declines in population growth rates by the high death rates associated with the brutal Amin and Obote dictatorships. Between 1970 and 1979, approximately 500,000 Ugandans lost their lives; from 1979 to 1986, another 300,000 perished as a result of warfare and human rights violations. Poor medical facilities, rampant diseases, growing emigration, and numerous economic problems also contributed to the decline in Uganda's population growth rate.

TABLE 4.1 Population Growth and Demographic Indices of the African Population Since 1911

	Population	Annual Growth Rate	Density/square mile
1911	2,463,469	—	—
1921	2,847,735	1.5%	—
1931	3,525,014	2.1%	61.2
1948	4,958,500	2.0%	34.9
1959	6,536,500	2.5%	46.0
1969	9,535,100	3.8%	67.0
1980	12,636,200	2.7%	88.8
1991	16,582,600	2.5%	116.4

NOTE: Population density figures are based on the total population, including non-Africans. Density percentages for 1931 are based on land and swamp; subsequent density percentages are based on land.

SOURCES: Republic of Uganda, *Background to the Budget, 1990–1991* (Kampala: Ministry of Planning and Economic Development, 1990), pp. 182–183; Republic of Uganda, *Background to the Budget, 1991–1992* (Kampala: Ministry of Planning and Economic Development, 1991), pp. 193–194; and *New Vision* (13 December 1990), pp. 6–7.

Despite this downturn in the population growth rate, Ugandan officials remained concerned about the population projections of the United Nations Fund for Population Activities (UNFPA). According to the UNFPA, by 2025, Uganda's population will be approximately 53 million. Such unchecked population growth would be hazardous to the country's social and economic development. To establish a national population policy, the Ugandan government and the UNFPA devised a Country Population Program for the 1988–1992 period. This $16 million program covered ten projects to control population growth and to improve the quality of life.[3]

In March 1992, the Ugandan government and the UNFPA sponsored a conference to discuss the problems associated with the country's high population growth rate. Among other things, the conferees established a committee to draft a national population policy. The committee's terms of reference included establishing realistic and timely demographic goals, determining ways to reduce fertility and improve the quality of life, and suggesting ways to reduce mortality rates. After drafting the policy, the committee will present it to the government for review and approval.[4] On 12 July 1993, the UNFPA announced that it had approved a US$14 million program for population awareness in Uganda. In part, the Ugandan government used the money to facilitate the adoption of a national population policy. By the end of 1994, there was no indication when Uganda would implement a national population policy.

Although commendable, these efforts will fail to control Uganda's population growth, despite the looming threat of the AIDS epidemic, which most experts predict will significantly increase the death rate. Cultural and economic inducements will remain the major reasons to have large families. Also, as more and more of Uganda is stabilized, international aid will increase, which will allow Kampala to devote more resources to improving the health care system. This will lower infant mortality rates and crude death rates. The effect of these developments will be continued population growth despite a slowing population growth rate.

Throughout the past century, Uganda's demographic patterns have changed at least three times. Historically, most Ugandans lived on agriculturally productive lands on the central plateau, which is about 200 miles from north to south and 150 miles wide. The second change occurred during the colonial period, when many Ugandans moved to cities and towns to take advantage of increased employment opportunities. The last change occurred between 1971 and 1986, when warfare and the lack of economic activity in urban areas forced many people back to traditional farmlands. As security and economic conditions improved, Uganda's cities and towns, especially Kampala, Entebbe, Jinja, Mbale, and Mbarara, have registered higher growth rates. Government demographers expect this trend to continue for the foreseeable future.

Language and Ethnicity

There are approximately thirty-two languages used in Uganda. English, Swahili, and Luganda are those most commonly spoken.[5] English is the official language and is used in government, commerce, academia, and the media. However, less than 30 percent of the population understand English, and most Ugandans who speak English are concentrated in urban areas. In rural areas, people normally speak an indigenous language, although many know a European language and some speak Arabic. The three major indigenous language families in Uganda are Bantu, central Sudanic, and Nilotic. Central Uganda's Lake Kyoga is the approximate boundary between the Bantu-speaking south and the northern central Sudanic and Nilotic language speakers. Bantu-language speakers are further classified into eastern Lacustrine, which includes the Baganda, Basoga, and many smaller groups, and western Lacustrine, which consists of the Banyoro, Batoro, Banyankole, and several lesser communities. Central Sudanic languages are spoken by the Lugbara, Madi, and some groups in northwest Uganda. The Nilotic language is subdivided into eastern speakers such as the Iteso, Karamojong, Jie, Dodoth, and Kakwa and western speakers such as the Acholi, Langi, and Alur.

In recent years, language has become more of a political issue for the government. During the 1970s, Idi Amin tried to foster national unity by extending the use of Swahili. At the 1990 session of the Organisation for Social Science Research in Eastern Africa in Kampala, President Museveni recommended the use of Swahili in eastern and central Africa to further inter-African integration. Other influential Ugandans have argued that the government should adopt Swahili as a national language and as the second official language in government. Advocates claim that Swahili would unify Uganda whereas opponents maintain that more Ugandans should learn English, the language of international commerce and politics, to help develop the country.[6]

There is considerable controversy about the role of ethnicity in contemporary Uganda. Some scholars maintain that British imperialists divided Ugandan society by developing the agricultural resources of the Bantu-speaking south and leaving the Nilotic-speaking north as a labor reservoir.[7] According to this view, the only people to benefit from these policies were the British and the Baganda, who willingly acted as British agents. Advocates of this theory also claim that all of Uganda's current ethnic problems can be traced to the colonial period. This interpretation of ethnicity, however popular, is incorrect. Many non-Baganda and non-Bantu-speaking groups prospered under British colonial rule. A typical example is the Iteso, a Nilotic-speaking people who live in an area from Karamoja to Lake Kyoga and grew wealthy after the 1912 introduction of cotton and coffee in their homeland.

Although all postindependence governments have downplayed the role of ethnicity, Ugandans continue to regard their family background as a critically important factor in their professional and private lives. Moreover, bureaucrats, businesspeople, farmers, and all other Ugandans rely on ethnic considerations to conduct their day-to-day affairs. Ethnic diversity has also thwarted all government plans for national integration.

Understanding the historical evolution of ethnicity in Uganda requires an appreciation of the role of the former kingdom of Buganda. Throughout the precolonial, colonial, and postindependence era, the Baganda have jealously guarded what they believe is their unique identity. On 31 December 1960, Buganda even declared its independence, an action the British promptly rejected as fatuous.

Since independence, every Ugandan government has sought the support of Buganda, the most influential and wealthy province in the country. The current Ugandan president, Yoweri Museveni, ingratiated himself with the Baganda by sanctioning the return and the coronation of Ronald Muwenda Mutebi II as the thirty-sixth *kabaka* of Buganda.[8] Needless to say, some Ugandans resented what they perceived as special treatment for Buganda. It is for this reason that repeated attempts to foster national unity have failed and that Uganda has stayed a divided country. It remains to be seen whether any Ugandan leader can convince all Ugandans that the leader can resist "the pressure from any one group while working successfully for the prosperity of all."[9]

Religion

Historically, religion has played a significant role in the national life of Uganda. Religious identification often influenced an individual's politics, employment, education, or social standing. Apart from a variety of traditional religions, Christianity and Islam are practiced throughout the country. Approximately two-thirds of the population are Christian (divided almost equally between Protestants and Roman Catholics) and about 15 percent are Muslim. The remaining 19 percent believe in traditional religions or do not practice any religion.[10]

Traditional religions are concentrated mainly in rural areas. Although Uganda's approximately forty ethnic groups have different traditional religious practices, all believe in one supreme being, except for Baganda traditionalists, who believe in several gods. The Baganda also worship spirits and ghosts of real and mythical ancestors. The Jie, Karamojong, and Dodoth do not practice ancestor worship. Nature spirits, which can take the form of animals, places, or natural phenomena, are also found in all traditional religions. Witchcraft and sorcery are present in many areas.

Christianity is Uganda's newest religion; it was introduced by European missionaries in the late 1800s. These missionaries represented several religious organizations, including the Anglican Church Missionary Society (CMS), the Roman Catholic Society of Missionaries of Africa (White Fathers), and Mill Hill Fathers. In the years leading up to the declaration of a British protectorate in 1894, religious wars between Catholics, Protestants, and Muslims plagued Buganda and other parts of southern Uganda. British and German military officers recruited Protestant and Catholic converts into their armies, which fought against one another and against Muslim forces. Although the Protestants and Catholics eventually agreed to stop fighting among themselves and to concentrate their activities in different regions of Buganda, religious competition has continued to divide Uganda.[11]

During the colonial period, British religious policy reflected a Christian bias. (Some Catholics maintained that the colonial government discriminated in favor of Protestants.) However, government officials also displayed a remarkable degree of "tolerance and neutrality" toward African traditional religions and Islam.[12] Nevertheless, Ugandan society continued to suffer from religious divisions, which soon entered the political arena. In November 1961, for example, Baganda Protestants created the Kabaka Yekka (KY) political party to demand Buganda's autonomy at independence. The Democratic Party (DP), which had been established in 1956 with the assistance of Catholic bishops, opposed the KY and Bugandan autonomy. The DP also gained many supporters in regions where there was considerable opposition to Buganda.

After independence, the issue of religion became increasingly important to Ugandan politicians. In August 1969, President Milton Obote, a Protestant, hosted the visit of Pope Paul VI, in part to demonstrate his denominational impartiality. Since becoming president in 1986, Yoweri Museveni has tried to lessen

Catholic-Protestant tension by refusing to be identified too closely with either religion. He has also urged the clergy to be nonpartisan and to keep out of politics. Nevertheless, religious leaders have remained an important factor in national politics. In August 1990, for example, Catholic priests who lived at Ggaba Seminary issued a statement imploring the government to allow multiparty politics. Additionally, various clergymen have stood for political office against the advice of their churches or have placed themselves in danger by providing support to one of several antigovernment guerrilla groups.[13]

Islam arrived in Uganda in the mid-nineteenth century from the East African coast via Arab and Swahili slavers and ivory traders. Muslim soldiers also brought their religious beliefs to Uganda from Sudan through the Nile River Valley. The majority of Muslims live in southern Uganda. Non-African Muslims (i.e., Pakistanis and Indians) usually followed some sect of the Shia branch of Islam whereas African and Arab Muslims belonged to the Sunni branch.

During the colonial period, factionalism plagued Uganda's Muslims. After the Grand Mufti, Sheikh Mbogo, died in 1921, the Muslim community divided into the Butambala and Kakungulu factions, largely because of disagreements over the selection of Mbogo's successor, who was to be the country's officially recognized interpreter of the Koran. In 1924, the colonial authorities appointed Haji Muhammed Ibrahim, an Islamic cleric from Tanganyika (now Tanzania), as a mediator between the two factions. However, he failed to resolve the dispute. In 1947, the Shaffii Mufti of Mecca also unsuccessfully tried to reunite Uganda's Muslims.

After ousting President Milton Obote in 1971, Idi Amin embraced Islam, primarily to gain support from radical Arab nations such as Libya. To preserve access to Tripoli's money and arms, Amin declared every Friday a public holiday to allow Muslims to worship. He also banned certain Christian sects and confiscated the businesses of many Christians and turned them over to Muslims. Finally, at the 1975 Islamic Conference at Lahore, Pakistan, Amin declared that Uganda was a Muslim country. Despite his popularity with Libya, Uganda's Muslim leaders distanced themselves from Amin because of the 1972 decision to expel the Asian community and his increasing use of violence and terror against anyone who opposed his regime.

After Amin's downfall in 1979, many Muslims fell victim to Ugandans who sought revenge for the outrages committed by Amin and his henchmen. Particularly vulnerable were the Kakwa and Nubian ethnic groups who had supported Amin. Yusuf Lule, who was president from 1979 to 1980 and a Muslim, succeeded in persuading many Ugandans to abandon their hostility toward Muslims.

Since 1986, government relations with the country's Muslim community have experienced periods of tension. A leadership dispute between Sheikh Ibrahim Saad Luwemba and Sheikh Mohammed Ziwa Kizito, both of whom claimed to be the grand mufti, divided Uganda's Muslims. On 19 March 1991, the Ugandan Supreme Court ruled that Sheikh Ibrahim Saad Luwemba was the rightful grand mufti. Three days later, about 2,000 members of the Tabliq Youth, an Islamic fundamentalist group that opposed both sheikhs, rioted at the Old Kampala Mosque and then killed an assistant police inspector and three constables. This incident, coupled with the subsequent arrest of ten Tabliq Youth members on murder charges, caused a gulf between the government and the Muslim community. On 1 November 1991, there were further problems when President Museveni accused an unnamed Muslim group of plotting to assassinate him. A few days later, Museveni cautioned Libya, Iran, and Saudi Arabia against trying to manipulate or disrupt "my Muslims."[14]

Since these two incidents, the Ugandan government has become more concerned about what it perceives as the growth of radical Islamic fundamentalism throughout the country. On several occasions, the authorities have banned public Muslim prayer meetings because they feared a breach of public peace. On 15 December 1993, Ugandan police arrested forty-six Tabliq Youth members who had refused to evacuate a building they illegally had turned into a mosque. According to some government spokesmen, the Tabliq Youth acted at the behest of Iran. Approximately two weeks later, the Ugandan government expelled Professor Syed Sufderul Huq, a Muslim academic, for encouraging the spread of Islamic fundamentalism.

Given the political fragility of Ugandan society and the growing activism of the country's Muslim community, the government undoubtedly will continue to perceive radical Islam as a threat. However, chances of a Libyan- or Iranian-financed Muslim revolution are minimal, largely because neither Tripoli nor Tehran has any strategic interest in Uganda. What is more likely to occur is that these two nations will provide modest support to certain Muslim individuals and groups in hopes of exacerbating existing tension between the Ugandan government and the Muslim community. A more immediate fear concerns Sudan's commitment to spreading radical Islamic fundamentalism throughout East Africa. As far as Kampala is concerned, the greatest threat emanates from Khartoum's support of the LRA, which is active in parts of northern Uganda, and of militant Muslims throughout Uganda. By late 1994, there were also unconfirmed reports that Sudan had given aid to fanatical groups such as the Uganda Islamic Revolutionary Front.

Education and Social Mobility

Traditional, Muslim, and European educational systems have existed in Uganda for centuries. Each possesses unique characteristics that have contributed to the intellectual enrichment of Uganda. Traditional education concentrated on teaching oral traditions and survival skills. However, traditional education, which was inward looking, failed to take cognizance of the world beyond East Africa. Muslim education, even though it focused on religion, represented a link to that greater global society. European educators also stressed the importance of subjects such as science and issues such as the immortality of man's soul, which were initially alien to many Ugandans. However, a European education offered an unprecedented degree of social mobility. As a result, an increasing number of Ugandans sought to attend European schools, which proliferated throughout the country. A more detailed examination of these educational systems will provide an understanding of the contributions each made to the evolution of Ugandan society.

Long before the coming of the Europeans, Ugandans had developed their own educational system. Teaching emphasized "learning by doing" and the creation of a sense of community and fellowship among students. Teachers also passed on oral traditions that provided students with an understanding of the trials, tribulations, and achievements of their ancestors. Most important, traditional education communicated the skills and knowledge required to live in regions that were often plagued by disease, warfare, drought, famine, and pestilence. In other words, children were taught how to adapt to their physical surroundings and how to survive in the face of adversity. Although it was invaluable for teaching Ugandans the skills necessary to survive, the traditional educational system failed to provide the country's population with an understanding of the larger world. To gain such knowledge, Ugandans relied on Muslim and European teachers.

Arab traders from Zanzibar and the East African coast first brought literacy to Uganda through their teaching of the Koran. Since 1844, Arab caravans had visited the court of Kabaka Suna. Apart from trading in guns, clothes, and other articles, the Arabs sought to convert the *kabaka* and his followers. Ahmed bin Ibrahim El Ameri, the first Arab to reach Buganda, therefore provided Koranic instruction to the *kabaka* and members of the royal court.[15] Although Arab conversion efforts were not very successful, Buganda adopted the Islamic calendar and many Baganda spoke Arabic.

Initially, European explorers competed with the Arabs for the religious loyalties of Ugandans. In 1862, John Hanning Speke presented a Bible to the *omukama* of Bunyoro-Kitara. After his arrival in Buganda in 1875, Henry Morton Stanley gave Bible lessons to Kabaka Mutesa. Two years later, eight CMS missionaries arrived in Buganda and established the foundation of a Western educational system that would eventually facilitate significant changes in Ugandan society and would enable successful students to enjoy a degree of social mobility unknown to their parents. Apart from elementary reading and writing, the missionaries taught basic arithmetic, history, geography, and offered instruction in manual or agricultural skills.[16] Even during this early period, Western education held out the prospect of social mobility. Dallington Maftaa, an English-speaking ex-slave from the coast who studied under Stanley, eventually became Mutesa's chaplain-secretary and acted as interlocutor between Buganda's royal court and the British. For his services, he received a chieftainship from Mutesa. Dallington Maftaa's educational experience served as an example to other Ugandans who wanted to improve their lives.

*Makerere University, the keystone of higher education in Uganda. Despite considerable aid
from various Western governments, the university continues to suffer from a lack of resources.
Nevertheless, the Ugandan government remains determined to restore Makerere to its former
grandeur. (Photo by Thomas P. Ofcansky.)*

After Uganda became a British protectorate, African education remained almost entirely in the hands of religious agencies (i.e., CMS, White Fathers, Mill
Hill Fathers, and Verona Fathers) under government direction. By 1921, more
than 152,000 Ugandans were enrolled in mission schools, the majority of which
were in Buganda and the Kampala area. Three years later, the Phelps-Stokes
Commission visited and criticized this uneven distribution. Moreover, the commission reported that the mission schools had failed to relate their educational
activities to the needs of the people, largely because they concentrated on literary
rather than technical training. Over the next few decades, Uganda's educational
infrastructure gradually expanded, and by 1951 the number of African students
enrolled in primary schools numbered about 180,000 and those in secondary
schools totaled approximately 5,500. There were also a number of schools for
Asian and European students. Higher education revolved around Makerere
University College, which attracted students from all over Africa. Despite these
achievements, numerous problems plagued the country's educational system.
African primary education suffered from a lack of teachers. Also, rival missions
often duplicated services, which inevitably meant that many schools had unoccu-
pied places. On a broader level, most Ugandans believed a Western education
would enable them to escape from the drudgery of an agricultural existence into

TABLE 4.2   Student Enrollment Between 1980 and 1991

Education Level	1980 Enrollment	1991 Enrollment
Primary schools	1,297,377	2,539,549
Secondary schools	73,093	223,245
Technical schools	3,457	8,579
Teacher training colleges	10,027	15,980
National teacher colleges	525	4,178
Technical colleges	789	N/A
Commercial colleges	624	2,490
University	3,916	7,468

SOURCES: James Tumusiime (ed.), *Uganda 30 Years 1962–1992* (Kampala: Fountain Publishers [1993]), p. 110; and Ministry of Planning and Economic Development, *Background to the Budget, 1991–1992* (Kampala: Ministry of Planning and Economic Development, 1991), p. 226.

white-collar positions. As a result, there was a bias against technical education, especially in the area of agriculture, which employed nearly 90 percent of the population. Nevertheless, by independence, Uganda possessed the most advanced educational system in East Africa.[17]

After 1962, the Ugandan government committed itself to making education more relevant to the country's needs. To achieve this goal, the authorities wanted to increase the number of technical schools, which alarmed many Ugandans who believed that such a policy would hamper social mobility. The Ugandan government also promised to provide every child with an opportunity to get an education.

Unfortunately, the country never achieved this goal, largely because the political instability of the 1970s and early 1980s prevented an expansion of the education sector. However, despite the absence of universal educational opportunities, the number of students in primary and secondary schools and in the universities and colleges more than doubled between 1969 and 1979. (Table 4.2 summarizes student enrollment between 1980 and 1991.) Despite efforts to rehabilitate the educational system in the early 1980s, continued economic problems reduced the quality of education at all levels.[18]

With the advent of his regime in 1986, Yoweri Museveni promised to improve and expand the country's educational infrastructure, eliminate illiteracy, reduce the teacher shortage and the high dropout rate, and narrow the gap between the high number of graduates and the low number of employment opportunities.[19] Long-term projects included establishing universal free primary education by the year 2002–2003, extending primary school from seven to eight or nine years, and shifting the emphasis in postsecondary education from academic to technical and vocational training.[20]

To achieve these goals, the Ugandan government formed an Education Policy Review Commission (EPRC). In 1989, the commission released its report, which accepted the premise that education would play a vital role in establishing national integration. The EPRC's recommendations stressed more equitable access to the educational system; improving the quality of instruction; and greater emphasis on learning practical skills in school, which would make graduates more competitive in the workplace.[21]

A government white paper subsequently outlined a strategy to implement the EPRC's findings. Among other things, the document suggested the creation of a National Council for Higher Education (NCHE). This organization would include a Joint Admissions Board for universities and other institutions of higher education, an examination board for schools other than universities, a Bureau of Academic and Professional Standards, and a Board for Human Resource Development. Additionally, the white paper urged that two new national universities be established, one in the north and one in the east. Numerous other recommendations called attention to the need for improved educational facilities and higher teacher salaries.[22]

President Museveni accepted these recommendations and quickly acted on many of them. By the early 1990s, as a result of rehabilitation and expansion efforts, the country's educational system included about 8,000 primary schools with approximately 2.3 million students. There were more than 500 secondary schools with some 230,000 students. About 15,000 students attended 92 primary teacher training schools and approximately 11,000 students studied at 10 teacher training colleges and 60 technical schools and colleges. Six commercial colleges enrolled about 1,000 students. Uganda also operated several universities, including Makerere University, Mbarara University of Science and Technology, Mbale Islamic University, the Uganda Martyrs University, and the Christian University of East Africa.[23] However, instability in many districts, especially Kumi, Apac, Gulu, Kitgum, Lira, Karamoja, Soroti, and Moyo, prevented numerous schools from remaining open.

In terms of the white paper's other recommendations, as of April 1992 the government had raised the teacher salary structure by 40 percent, which will form the basis of future general salary increases. Additionally, officials arranged to provide all teachers with a professional allowance and to streamline the teacher promotion process. By late 1993, the NCHE was supposedly operational. Plans also went ahead for the development of two new national universities, one of which will be developed as a polytechnic. Most important, the Ministry of Education drafted a five-year (1992–1993 to 1996–1997) Shs. 264 billion educational investment plan. After completion of this phase, the total additional recurrent expenditure is estimated to be Shs. 786 billion.[24]

The success of Uganda's educational policy will depend on several factors. First, Kampala must secure adequate foreign assistance to finance the reforms. Although it received a $108 million U.S. grant for a ten-year primary education improvement program, the Ugandan government will need sustained infusions of foreign economic and technical aid to rehabilitate the educational system.[25] Next, there will have to be significant improvements in Uganda's economy if the dream of social mobility via education is to become a reality. Last, the Museveni regime will have to establish stability in its border regions if all the country's schools are to operate on a regular basis.

Health

Throughout its history, Uganda has suffered from numerous diseases, including sleeping sickness; malaria; tuberculosis; cerebrospinal meningitis; smallpox; diarrhea; and, more recently, the AIDS epidemic.[26] During the precolonial period, Ugandans relied on traditional healers and medicines to cure their ills. By introducing Western medicine in the 1890s, the British altered the centuries-old balance between high birthrates and high death rates.[27]

The construction of hospitals, the establishment of preventive medical services and research facilities, and the relocation of large numbers of Africans away from tsetse fly–infested regions to protect them from sleeping sickness helped to precipitate a population explosion. Controlling diseases was not the only reason for Uganda's population growth. The British also reduced tribal warfare and introduced modern agricultural methods. As a result, from a 1903 estimate of 3.5 million, Uganda's population more than doubled to about 7.2 million at independence.[28]

Early medical work concentrated on controlling the spread of various diseases, the most serious of which was sleeping sickness. Between 1901 and 1905, this disease had claimed at least 200,000 Ugandans who lived along Lake Victoria or on the Buvuma Islands. In 1906, Uganda's governor, Sir Hesketh Bell, accepted the recommendation of his medical advisers and resorted to "drastic measures" to relocate Africans from infected areas to government-operated camps.[29]

Although it failed to eradicate sleeping sickness, this strategy achieved several objectives. The authorities succeeded in evacuating Ugandans from high-risk zones in the Lake Victoria region. In the camps, patients received greater care than they would have had they remained at home. Gradually, the sleeping sickness epidemic lost its morbid mystique and became one of the many maladies that posed threats to Uganda. The colonial government therefore could devote more time and resources to sanitation, preventive medicine, and health education.

World War I had a significant impact on Uganda's medical history. In August 1914, the British decided to create an East African medical corps, whose ranks would be open to Africans. By 1915, the medical authorities had established and trained the 1,000-man Uganda Bearer Corps, which later was incorporated into the African Native Medical Corps.[30] This unit, which saw service throughout East Africa, earned the admiration and respect of Uganda's medical establishment.

After the war, Major G. J. Keane, who had organized and commanded the African Native Medical Corps, urged the colonial authorities to provide medical training for corps veterans and other qualified Africans and to form a Civil Native Medical Service. Although most officials supported these ideas, Keane had to work many years before he succeeded in persuading the colonial government to provide medical training to Africans.

In 1921, the efforts of Keane and other colonial officials resulted in the opening of Makerere College with fourteen African students, who received technical instruction. The following year, the college changed its curriculum and started offering medical courses, which had been taught at Mengo Hospital by the CMS. During the first two years, students received basic medical training at Makerere and for the last three years studied physiology, anatomy, materia medica, surgery, medicine, and other subjects at Mulago Hospital. It should also be pointed out that the CMS played an essential role in Uganda's early medical history. On 22 February 1897, a small group of missionaries, which included Albert R. Cook, started seeing patients at Mengo Hospital, which at that time was little more than a primitive bush dispensary. During the following year, more than 17,000 Africans sought medical care at Mengo.[31] In 1927, the first Africans qualified as senior African medical assistants.

During its brief time as a British colony, Uganda developed one of the finest Western medical systems in sub-Saharan Africa. Unfortunately, Idi Amin and his soldiers destroyed this infrastructure. A massive exodus of Ugandan, Indian, and European doctors, nurses, and medical technicians created severe personnel shortages throughout the health care sector. Fighting also destroyed many hospitals and clinics. The decline in medical services caused higher infant mortality rates and reduced life expectancy.[32] Until 1986, a succession of governments did little to undo this damage.

Since then, the situation has gradually improved. By 1991, the Ugandan government—assisted by parastatals, private practitioners, and numerous nongovernmental organizations (NGOs)—had restored a health care system in urban areas and near trading centers. This system consisted of 81 hospitals, 48 of which belonged to the government; 105 health centers (97 government operated), and 765 lower health units (601 government operated).[33] Rehabilitation of rural health service remained one of the government's top priorities.

Apart from starting to rebuild the medical infrastructure, the Museveni regime has succeeded in improving the country's life expectancy, crude death, and infant mortality rates. Despite these advances, Uganda still faces a debilitating array of diseases. According to Ugandan health officials, measles, respiratory tract infections, and gastroenteritis accounted for one-half of all illness-related deaths. Other fatal illnesses included anemia, tetanus, malaria, tuberculosis, whooping cough, and diarrhea.[34] But the AIDS epidemic overshadowed all other medical concerns and posed the greatest health threat to Uganda.

In late 1982, the first reported AIDS-related deaths occurred at Kasensero port in Rakai District, when seventeen traders died from the disease. Rapidly, AIDS seemed to spread to Rakai Town, Masaka, along the trans-African highway to

Mulago Hospital, Kampala. Despite the rehabilitation of many Ugandan hospitals, health care remains woefully inadequate for most Ugandans. Resolving this problem is one of Uganda's top priorities. (Photo by Thomas P. Ofcansky.)

Kampala, and to numerous towns throughout eastern Uganda. By the early 1990s, medical officials acknowledged that AIDS had been reported in "virtually all districts." Since 1982, the number of AIDS cases has been doubling every six months. By 1993, one out of four Ugandans had AIDS, according to Dr. Benon Biryahwaho, a staff member of the Uganda Virus Research Institute in Entebbe. He also indicated that more than 9 percent of the children less than eighteen months old have AIDS (acquired prior to birth or through breast-feeding). Additionally, Dr. Biryahwaho claimed that with almost 15 percent of pregnant women infected with the disease, the mother-to-child AIDS transmission rate was 60 percent, which was far higher than the generally expected rate of 35 to 50 percent. These shocking statistics caused Dr. Elizabeth Madra, head of Uganda's AIDS Control Program (ACP), to warn that "the nation is in danger of losing most of its people."[35]

For a desperately poor country like Uganda, the AIDS epidemic is a catastrophe of frightening proportions. Years of guerrilla warfare and social unrest, which all but destroyed the country's health care system, coupled with initial official silence about the disease worsened the problem. To its credit, however, Uganda was one of the first nations in sub-Saharan Africa to acknowledge the presence of AIDS. Moreover, by the early 1990s, the country had adopted one of the region's most aggressive anti-AIDS campaigns.

In 1986, for example, the government established the ACP, which is a special Ministry of Health unit. The ACP has twelve professional staff members, assisted by five World Health Organization (WHO) specialists, and focuses on health education, safe blood supply (laboratory and blood transfusion services), monitoring the epidemic, and patient care. The ACP also helped to rehabilitate the Uganda Virus Research Institute. In the districts, the ACP has recruited and trained a cadre of local health care workers who report to district medical officers. WHO and United Nations Children's Fund (UNICEF) have provided these personnel with three Suzukis and thirty-nine motorcycles for transport and funds for operating costs. At the village level, resistance councils, community leaders, churches, and schools have devised a variety of anti-AIDS programs.[36] Another ACP activity concerned the UNICEF-sponsored School Health Emergency AIDS Program. By mid-1989, this program was active in thirty out of thirty-three districts with 2,400 science teachers participating in workshops. The ACP has also certified 2,450 primary school teachers and 100 secondary school teachers as trainers. These personnel taught about 200,000 students in approximately 600 secondary schools about the dangers of AIDS. Additionally, for schools they launched a new science syllabus that included warnings about AIDS and distributed about 20,000 anti-AIDS school kits.

On a broader level, the Ugandan government established AIDS clinics in Mulago, Nsambya, Rubaga, and Kitovu and installed blood-testing machines at twenty-eight screening centers throughout Uganda. In February 1990, an AIDS information center opened in Kampala, which offered free blood testing and pretest and posttest counseling to the general public. Additionally, the Ugandan government has sponsored scores of seminars and workshops for government and military personnel, journalists, prisoners, university students, health care workers, and clergy. Health officials have distributed almost 7 million pamphlets and four tons of anti-AIDS badges, posters, and stickers in ten languages.[37]

There have also been scores of other public and private initiatives. Many of Uganda's churches have taken up the cause of AIDS education. Numerous clergymen, including Anglican bishop Misaeri Kauma, mentioned the dangers of the disease every time they preached. Some churches printed leaflets and posters and distributed them at services and in free Bibles.[38]

As it became clear that Uganda's health care system could not cope with the AIDS epidemic, the authorities sought international assistance to combat the disease. In 1992—with support from the World Bank, UNICEF, and the United States Agency for International Development (USAID)—President Museveni authorized the creation of the Uganda AIDS Commission. In November 1993, this organization launched a five-year plan to control the spread of the human immunodeficiency virus (HIV), the virus that causes AIDS, among children and adults, especially women. This plan, which aimed to educate the country's population about the dangers of AIDS, received more than $125 million from the international donor community and about $37 million from the Ugandan government. The commission also arranged to test various AIDS vaccines on Ugandan volunteers and worked with the country's herbalists through their organization Uganda N'eddaggala Lyayo in searching for a cure to AIDS.[39]

Numerous NGOs have also helped Uganda combat the ramifications of the AIDS epidemic. World Vision, a California-based Christian relief and development organization, has worked with the Ugandan War Widows Foundation to help orphans in Masindi, Rakai, and Gulu. The major goal of the two groups is to place orphans in foster homes and to give tuition support so that the children can attend public school. Other NGOs that are working on anti-AIDS programs in Uganda include Save the Children, Action Aid, the Experiment in International Living, and the AIDS Support Organization.[40]

Notwithstanding these encouraging developments, the AIDS epidemic has had a severe impact on Uganda. The disease, which has caused tens of thousands of deaths and has generated at least 120,000 orphans, has affected all social sectors. High infection rates among young, educated Ugandans threaten not only to slow the pace of economic development but also to deprive Uganda of the human resources needed to resolve Uganda's many political, social, and economic problems. As a result, Uganda faces an increasingly bleak future. According to a 1994 study by a group of international medical researchers, the AIDS epidemic will continue to grow at an "extraordinary rate of growth," despite Uganda's extensive health campaigns.[41]

The Role of Women in Modern Uganda

Historically, women have played a significant role in Ugandan society. Apart from performing the traditional duties associated with child rearing, women have formed the backbone of the country's cash crop agricultural sector. In addition, Ugandan women have enjoyed considerable status in many traditional societies. Among other things, women have often served as religious leaders who sometimes led revolts against the established order. In parts of Uganda, women have owned land, participated in the political process, and cultivated crops for personal profit.

During the colonial era, European missionaries sought to improve the status of women by agitating against the brutal practice of clitoridectomy, opposing polygamy, and instituting Western educational courses for women. Some African communities and the colonial government supported these efforts. In 1919, for example, the colonial government banned Baganda native courts from recognizing polygamous marriages. During the same year, the secretary of the Young Baganda Association argued that women had to receive "a good education" to facilitate Uganda's development.[42] A 1925 white paper also acknowledged the importance of women's education, and a Royal Commission on Girls' Education urged that women receive advanced agricultural training. Such efforts enabled Uganda to institute a Western educational system for women that was superior to those in Kenya and Tanganyika (now Tanzania). After receiving a Western educa-

Most Ugandan women pursue a traditional way of life. However, in recent years women have become more politically active and have played a prominent role in Yoweri Museveni's government. (Photo courtesy of U.S. Committee for Refugees.)

tion, most women worked in the teaching and nursing fields. However, as Uganda became more industrialized, employment opportunities for women opened in shops, offices, farming, nursery care, dressmaking, designing, tailoring, and book-binding. Advances on the national level increased after the colonial government sanctioned the 1947 establishment of the Ugandan National Council of Women, which organized clubs throughout the country to "safeguard and improve family life, encourage women to understand and respond to the demands of citizenship in modern African society."[43]

Despite these changes, by the beginning of the 1960s little had changed for the millions of Ugandan women who lived in rural areas. Husbands normally appropriated money earned by their wives. Traditional African societies also severely restricted women's personal freedom and many encouraged female circumcision. According to one account, these practices existed because when a woman "sees her mother and all the women around her accepting [a] position of inferiority, she accepts it too."[44]

During the final days of colonial rule, women's issues became more visible. The Ugandan National Council of Women, in conjunction with the Ministry of Information, issued a series of citizenship leaflets. The first of these argued for equal opportunities for education and public office, better medical services for women, and an end to discriminatory laws. Additionally, there was a "Women Look Ahead Conference," which examined ways to improve the economic well-being of women. During this period, women also became more politically active. In the February 1962 preindependence elections, for example, voters elected Sugra Vishram, Florence Lubega, and Eseza Makumbi to Buganda's *lukiko*.[45]

This momentum continued through the early years of independence. The Ugandan National Council of Women advocated legal reforms that would allow women to own property and to retain custody of their children if their marriages failed.[46] However, this activism slowly dissipated as women's issues became less and less of a national priority. There was also little progress in the political arena. During Milton Obote's first government, there were only three women members of parliament and no women ministers. Additionally, Milton Obote's *The Common Man's Charter*, which heralded Uganda's "move to the left," failed to offer any solutions to the problems facing Ugandan women. This state of affairs was due, at least partially, to the reticence of women themselves. The president of the Ugandan National Council of Women, Florence Lubega, reflected this attitude when she welcomed Milton Obote to a council meeting by saying she had no intention of organizing "a rebellion against men."[47]

From 1971 until 1986, all Ugandans suffered horribly as a result of violence, instability, and warfare.[48] For women, these years reversed many of the gains of the past several decades. In 1973, the government banned all women's organizations; between 1983 and 1985, the authorities opposed attempts by the Ugandan National Council of Women to improve the status of Ugandan women.[49] More important, widespread chaos all but destroyed normal family life. The reduction of public services and the devastation of the country's infrastructure reduced access to schools, hospitals, and markets. As a result, employment opportunities waned, forcing many women to revert to subsistence agriculture to live. Apart from these misfortunes, women suffered sexual abuse and countless other humiliations at the hands of soldiers, rebels, and brigands. One of the most pernicious indignities occurred during the Amin regime, when soldiers accused all unmarried women of being prostitutes. The government then ordered single women to be cleared from cities and towns. Senior military officers also mounted a crusade to force all Ugandan women to marry.

After 1986, conditions for women slowly started to improve. In his first address to women, Yoweri Museveni argued that their status would change only within the context of nationwide progress for all Ugandans. Many Ugandan women, however, rejected his speech and Museveni's Ten-Point Program for not targeting women as a group to be given preferential treatment. Despite this opposition, Museveni quickly established a reputation of doing more than any postindependence Ugandan leader to advance women's rights. In 1988, for example, he established the Ministry of Women in Development, largely as a result of pressure by a group called Action for Development. The ministry formulates and implements women's programs, trains women in leadership skills, and sensitizes all Ugandans to women's issues. Additionally, the ministry has involved Ugandan women in the process of revising the 1967 Constitution.[50] In the same year, the Ugandan government elevated the Women's Desk in the NRM secretariat to the Directorate for Women's Affairs. Its role is similar to that of the Ministry of Women in Development, but the directorate is more active in political mobilization. Among other things, the directorate conducts seminars and workshops throughout Uganda to involve women in national issues.

Apart from these developments, President Museveni has appointed numerous women to senior government positions. In 1990, for example, there were eight women holding ministerial portfolios. The NRA also welcomed women to join its ranks. On the local level, many women serve as RC leaders or hold other important grass-roots posts.[51]

This progress has failed to change basic attitudes toward women, especially in the rural areas. Polygamy is widely practiced, and wife beating is an accepted practice. According to the Uganda Women's Lawyers Association (UWLA), an estimated 50 percent of marriages in Uganda are polygamous. In 1991, UWLA embarked on a program to convince women that wife beating is not a sign of a man's love. Within many families, boys are encouraged to attend school; girls are not. Widows are regularly displaced from their matrimonial homes as a result of unfair inheritance and succession laws. According to many female activists, the early 1990s have witnessed a dramatic increase in sexual abuse throughout Uganda, largely because the courts were lax in punishing sex offenders.[52] To resolve this problem, women representatives in the NRC succeeded in introducing various amendments to the penal code that required stiffer sentences for those found guilty of sex crimes. However, prospects for a significant reduction in the number of sexual offenses are not encouraging. Miria Matembe, who is director of Action for Development, believes that sexual harassment stems from "long standing ex-

A cultural treasure: The National Theatre, Kampala. After 1986, Ugandan culture quickly reestablished itself. Today, Ugandan art, music, and films are known throughout the world. (Photo by Thomas P. Ofcansky.)

ploitative and oppressive cultural practices," which will be extremely difficult to change.[53]

Popular Culture, the Arts, and Literature

Uganda has a rich and diverse cultural history. Apart from traditional arts and crafts such as wood carving, pottery, weaving, and jewelry making, music and dance have played an important role in the daily lives of all Ugandans. Also, oral literature has preserved the country's history and genealogies through myths, legends, and folktales. Storytellers enjoy considerable prestige in their communities.

The coming of the Europeans stimulated interest in literature and the publishing industry. Sir Apolo Kagwa, prime minister of Buganda from 1899 to 1926, wrote several books about the history and customs of the Baganda and their neighbors. In 1971, one of Kagwa's more important books was translated into English; it remains essential reading.[54] Other noted Ugandan writers include Ham Mukasa, Daudi Chwa II, E. S. Kironde, and William Kigongo. In January 1934, the *Uganda Journal* first appeared. Although its subscribers were primarily European, some Ugandans contributed articles to the journal, which, although no longer published, remains an important source of information on all aspects of Uganda's historical development. Another significant periodical, *Transition,* which was founded in October 1961 under the auspices of the International Association for

Kampala skyline (Photo by Thomas P. Ofcansky.)

Cultural Freedom, contained an array of articles on literature, politics, economics, and poetry. In October 1968, the government banned *Transition* and ordered the arrest of its editor on charges of publishing seditious material.

There were numerous other cultural activities in Uganda. The Uganda Society carried out work in the fields of history and literature. The Uganda Museum assembled a large collection of art, local crafts, and musical instruments. The National Theatre presented a variety of plays and musicals. The National Dance Troupe of Uganda performed throughout the country and abroad. Radio Uganda, founded in 1954, and the Uganda Television Service, established in 1963, broadcast domestic and international news and entertainment programs.

The violence and instability of the 1970s and early 1980s ended or severely restricted artistic expression. After 1986, Uganda slowly started to rehabilitate and rebuild its cultural infrastructure. Between 1986 and 1989, the Ugandan government, with the assistance of the United Nations and Canada, repaired the Uganda Museum and rescued many historical and cultural exhibits. Uganda also renovated the Adult Education and Cultural Center at Mengo, which now has an exchange program for artists and actors. Additionally, the center offers day courses and cultural programs for youths and various ethnic groups. By 1992, Makerere University's Margaret Trowell School of Fine Art was graduating more than forty students annually. Although many graduates find employment in the business sector as graphic artists, some become art teachers or artists.[55]

After years of little or no activity, the National Theatre has revived the performing arts. Among its many activities, the National Theatre is the venue for annual music and drama festivals and the headquarters for the Uganda Theatrical Groups Association. In 1990, Japan agreed to finance the renovation of the National Theatre.[56]

During the post-1986 period, Uganda developed a small film industry. In late 1990, international director and producer Mira Nair shot part of the film *Mississippi Masala* in Uganda. A documentary, called *Alone,* was made in Uganda, as was one drama about the AIDS epidemic, entitled *It's Not Easy.*[57]

Although still recovering from the excesses of the 1970s and early 1980s, Uganda Television Service has installed five relay stations, at Masaka, Mbale, Mbarara, Lira, and Soroti. Parts of northern and western Uganda are still awaiting the inauguration of television service. On 15 December 1992, Uganda launched its second television channel to broadcast news, documentaries, and other programs of the U.S. Cable News Network (CNN), which is beamed by Cablesat International. Initially, the service has been confined to the greater Kampala area, but there are plans to extend service to rural viewers as soon as necessary facilities are operational. In early 1993, Uganda signed an agreement with Stem Cable and Satellite Limited of South Africa to broadcast films to subscribers within a 25-mile radius of Kampala.[58] A Ugandan company, International Television Network Limited, also planned to provide programming throughout the country. Radio Uganda broadcasts about 230 hours a week in twenty-two languages, including English, French, Arabic, Swahili, and numerous local languages. On 22 February 1993, the Ugandan government broke its thirty-year monopoly of the airwaves by licensing a private station known as Radio Kampala. By late 1993, two other stations, Capital Radio and Sanyo Radio, had also received government licenses. All stations broadcast music and light entertainment, educational programs, and news reports within the greater Kampala area.

5

THE ECONOMY

W HEN IT GAINED INDEPENDENCE in 1962, Uganda had one of the most prosperous economies in Africa. The country, which possessed a good climate and fertile soil, was self-sufficient in food. Approximately 93 percent of the population worked in the agricultural sector, which was a source of considerable foreign exchange earnings. The manufacturing sector satisfied most of Uganda's consumer needs and also earned some foreign exchange through textile and copper exports. Export earnings financed the country's import requirements and resulted in a current account surplus. Uganda's transport infrastructure, which ranked among the best in Africa, included effective networks of roads, railways, and port and air transport. There was also a small wage-earning class, concentrated in Uganda's cities and towns, that was employed in the service sector. The Asian community dominated trade, small industries, and businesses.

During his first presidency (1962–1971), Milton Obote initially sought to preserve the economic system created by the British. In his election manifesto, Obote promised that the country's economy would remain based on the three Cs—coffee, cotton, and copper—supported by the three Ts—tea, tobacco, and tourism. However, as a result of fluctuating international coffee and cotton prices, Obote tried to reduce Uganda's dependence on the export-oriented cash crop economy. To accomplish this goal, he expanded the tourism and mining sectors and promoted indigenous industries.[1]

Increasing political pressure to end Uganda's dependence on the capitalist world market and to Africanize the economy prompted Obote to adopt a series of radical policies. To increase African participation in the wholesale and retail trade, the government established the African Business Promotions (1963) and the National Trading Corporation (1966). The 1969 Trade Licensing Act authorized the Ministry of Commerce to ensure that trade was controlled by firms whose shareholders were predominantly Ugandan.

To offset the negative impact of these acts on foreign investors, the Industrial Charter of 1964 guaranteed adequate compensation for nationalized properties, allowed foreign enterprises to import machinery and other materials duty free, and promised that foreign businesses would be treated the same way as Ugandan-owned firms. In May 1970, however, Obote alienated the foreign business community by issuing the Nakivubo Pronouncement, which sanctioned the 60 percent nationalization of eighty major foreign businesses operating in Uganda.

Obote's radicalism failed to bring about Uganda's economic self-reliance. In the mid-1960s, declining international demand forced a fall in coffee and cotton prices. Ordinarily, such a development would have caused widespread hardship throughout the country. However, Uganda survived this difficult period because of increased production of other crops. In 1966, for example, tea production increased by more than 30 percent to 24.7 million pounds as opposed to 18.4 million pounds the previous year. Additionally, there were increases in the sugar and tobacco crops. Significant growth in the manufacturing sector also helped Uganda maintain a favorable balance of payments.

As a result, gross domestic product (GDP) during the 1963–1970 period increased by 4.8 percent per annum, or about 2 percent per capita. In absolute terms, per capita income rose from $86 in 1963 to $134 in 1970. Additionally, export earnings were more than adequate to cover import requirements and a current account surplus was achieved in most years.

After he seized power in 1971, Idi Amin, like his predecessor, promised to place Ugandans in control of their own affairs and to accelerate the Africanization of the economy. To achieve these goals, Amin unleashed a so-called war of economic liberation. During the 1971–1974 period, this "war" resulted in the appropriation of wealth from almost the entire expatriate community, including Asians, Britons, Israelis, non-Ugandan Africans, foreign missionaries and teachers, and American Peace Corps personnel. Altogether, the authorities seized 5,655 businesses, factories, ranches, and farms. Additionally, after expelling the Asian community in 1972, the government confiscated their "abandoned" property (i.e., homes, vehicles, and household effects).

Amin gave at least 500 businesses to his friends and supporters. He then established the Business Allocation Committee of the Ministry of Commerce and ordered it to distribute the remaining spoils of the "war of economic liberation" to all Ugandans. In practice, however, these committees, which eventually came under control of the military, transferred most properties to soldiers and bureaucrats, few of whom had any business experience. Consequently, the commercial infrastructure that provided Ugandans with daily necessities such as soap, blankets, kerosene, cloth, matches, sugar, and salt quickly collapsed.

The "war of economic liberation" also crippled the rest of the economy. By late 1972, a deficit in Uganda's balance of payments reduced foreign reserves to an all-time low of Shs. 81 million, an amount sufficient for only three weeks' imports. Industrial production also suffered. Between 1971 and 1979, sugar production fell from 155,675 to 13,697 tons. During the 1971–1975 period, soap production dropped from 13,600 to 3,600 tons. Additionally, production of cement, pipes and sheets, steel ingots, corrugated iron roofing, matches, gunny bags and sacks, and blankets declined by anywhere from 55 to 94 percent. By 1978, copper mining and smelting and fertilizer production had ceased.

On a broader level, Amin's reign of terror destroyed the very foundations of Uganda's economy. Between 1970 and 1980, Uganda's GDP declined by approximately 25 percent, exports by 60 percent, and imports by almost 50 percent. Large defense expenditures further ravaged the economy. Amin financed the government budget by excessive bank borrowing, which resulted in an average annual inflation rate of more than 70 percent.

Regional trade also suffered as a result of the July 1977 breakup of the East African Community (EAC), a common services organization operated by Uganda, Kenya, and Tanzania. The EAC's collapse occurred because Kenya eventually dominated about 50 percent of all regional trade. As a result, Uganda and Tanzania complained that the EAC was little more than a front for Kenyan businessmen. Additionally, Tanzania, which had embraced a socialist economic policy, felt increasingly uncomfortable with Kenya's capitalist ideology.[2]

During his second presidency (1980–1985), Milton Obote succeeded in getting IMF assistance for an economic recovery program. During the 1980–1983 period, this initiative stimulated some economic growth in the agricultural sector but failed to generate any significant activity in the manufacturing sector. The lack of donor support and the IMF's decision to terminate its program in Uganda because of disagreements with Obote over high public spending rates and the government's budget policy prevented the country's economic rehabilitation. During Tito Okello's brief six-month government (1985–1986), the economy again plunged into chaos as a result of political instability and civil war.

The National Resistance Movement's Recovery Program

When he seized power in January 1986, Yoweri Museveni inherited a ruined economy. After being sworn in as president on 28 January 1986, he promised to rehabilitate the economy and to make a "fundamental change" in the lives of all Ugandans. Museveni also announced the implementation of a Ten-Point Program, which he said would pull the country out of its misery and build an independent, integrated, and self-sustaining national economy. The ten points are as follows.

1. Establish popular democracy.
2. Restore security.
3. Encourage national unity.
4. Maintain national independence and nonalignment.
5. Rebuild the economy.
   a. Diversify agriculture.
   b. Build industries in the import-substitution sectors.
   c. Encourage rapid industrialization.
   d. Reconstruct basic industries.
   e. Manufacture machine-making machines.
   f. Acquire computer technology.
   g. End dependence on other nations.
6. Restore and rehabilitate social services.
7. Eliminate corruption and misuse of power.
8. Resettle displaced people.
9. Promote regional cooperation and human rights.
10. Create a mixed economy.

To achieve these goals, the government devised a multiphase strategy whose main elements included diversification of agriculture, development of agro-based industries, building of import substitution industries, establishment of domestic capacity to produce capital goods, and improvement of the managerial infrastructure. In 1987, the Ugandan government launched the Rehabilitation and Development Plan (RDP) and the Economic Recovery Program (ERP).[3] Prior to the implementation of these two programs, in February 1986, the Ugandan government had adopted a Relief and Rehabilitation Plan to provide emergency assistance to the war-ravaged areas of Luwero and West Nile. Although it achieved only minimal results initially, this plan helped to generate some public confidence in the Museveni regime.

On 15 June 1987, the NRM unveiled the four-year RDP for fiscal years 1988–1991. (Table 5.1 reviews the RDP's financial goals.) The plan sought to revitalize Uganda's industrial and agricultural productive capabilities; rebuild the economic, transportation, and social infrastructures; reduce inflation by 10 percent annually; and stabilize the balance of payments.[4] The RDP required $1,289 million over the four-year period.

The RDP's results were mixed. Although the annual growth rate increased from 0.3 percent in 1986 to 6.1 percent in 1987 and 7.2 percent the following year, the program failed to cure the country's economic ills. Some of the major problems that hampered the RDP included an overvalued currency, high government spending, and an annual average inflation rate of almost 240 percent. Exports also declined by 18 percent in 1987 because of a drop in coffee sales and delays in

TABLE 5.1 Sector Allocations of the Rehabilitation and Development Plan for Fiscal Years 1988–1991 (in millions of U.S. dollars)

	Proposed Spending	Percent of Total	Funding, March 1988	Balance
Transportation/ communications	378.7	29.4	171.0	207.7
Agriculture	314.5	24.4	186.8	127.7
Industry/ tourism	271.9	21.1	93.7	178.2
Social infrastructure	221.0	17.2	90.8	130.2
Mining/energy	89.2	6.9	53.4	35.8
Public administration	13.2	1.0	1.2	12.0

SOURCES: The Economist Intelligence Unit, *Country Profile: Uganda, 1989–90* (London: Economist Intelligence Unit, 1989), p. 10; and Rita M. Byrnes (ed.), *Uganda: A Country Study* (Washington, D.C.: USGPO, 1992), p. 242.

donor assistance and in adjusting the exchange rate. Most important, the government's inability to end guerrilla wars in northern and northeastern Uganda impeded the country's economic rehabilitation and required and caused excessive defense expenditures.

At about the same time that it activated the RDP, Uganda implemented the ERP, whose goals included demonetizing the Uganda shilling by a factor of 100; applying a 30 percent conversion tax on all cash holdings, deposits, and treasury bills during the currency exchange to syphon off liquidity amounting to Shs. 2.69 billion; devaluing the Uganda shilling by 76.7 percent; increasing producer prices of export crops by a factor of three to five times; and adopting conservative fiscal and monetary policies and reducing government indebtedness to the banking system. On 24 July 1987, the Ugandan government announced numerous other policies, some of which had their origins in the Ten-Point Program, to support the ERP. Some of the more important initiatives included a 20 percent reduction in interest rates and an initiative to restore the properties of Asians who had been expelled from Uganda during Idi Amin's regime.[5]

The IMF and the Paris Club supported the ERP by agreeing to a Structural Adjustment Facility. This financial package included $185 million worth of debt rescheduling for 1986–1987, a $200 million economic recovery credit, and other foreign assistance. In July 1988, President Museveni implemented another stabilization program, which aimed to sustain economic growth and to reduce inflation. Many Ugandans opposed the IMF and the Paris Club for dictating economic policy. When the Structural Adjustment Facility failed to result in a quick economic recovery, many Ugandans criticized the NRM for implementing an economic strategy that, in their view, had been determined by foreign financiers (i.e., the IMF and the Paris Club). Nevertheless, President Museveni went ahead with the July 1988 program.[6]

The ERP's performance was mixed. In 1987, the GDP rose by 4.5 percent and the industrial and agricultural sectors grew by approximately 15 percent and 8.5 percent respectively. However, the annual average inflation rate remained at about 24 percent. Reasons for the ERP's poor showing included the lack of control over fiscal and budgetary policies, which stymied all efforts to rebuild the economy.

Apart from these initiatives, the Ugandan government devised at least two other programs to facilitate economic reconstruction. In January 1991, for example, the NRM signed into law an Investment Code that sought to facilitate and monitor foreign and domestic investment in Uganda. This code offered tax and other incentives to local and foreign investors and created the Uganda Investment Authority (UIA), under the Ministry of Planning and Economic Development. The code enabled investors to secure all required operating licenses and approvals from the UIA rather than from various government ministries and agencies. In addition to serving a regulatory role, the UIA encouraged foreigners to invest in Uganda and helped them identify and establish suitable investment projects. Within its first year of existence, the UIA compiled a fairly impressive record. According to UIA executive director George Rubagumbu, the UIA had received eighty-six investment applications worth $281 million by the end of February 1992 (foreign investors accounted for $223 million of this amount). The Ugandan government had approved thirty-six projects worth $120 million and had implemented twenty-one projects.[7]

Barter trade arrangements became one of the NRM's major tools because they supposedly helped Uganda save its meager foreign exchange reserves, diversify exports, encourage south-south cooperation, and reduce reliance on the West.[8] In 1986, Uganda started bartering surplus coffee with nations outside the International Coffee Organization (ICO) for a variety of commodities, ranging from pharmaceuticals and machinery to sugar, salt, petroleum products and tractors.[9] By 1992, the Ugandan government had committed itself to some thirty-five barter trade agreements worth an estimated $600 million. These agreements proved to be a bureaucratic nightmare in an economy geared for cash transactions. Only nine have been completed. Nineteen remain active, although all of them are behind schedule, and the government will probably cancel the rest. Moreover, barter trade proved to be a money loser because of a costing system that based prices on those current at delivery time rather than those in effect at the time the agreement was concluded. Since it normally bartered coffee, Uganda lost more and more money as coffee prices declined. Eventually, the government banned the use of coffee for any new barter trade agreements, even though none are likely anyway.[10]

TABLE 5.2 Major Economic Indicators, 1990–1992 (millions of U.S. dollars unless indicated)

	1990	1991	1992 (provisional)
Domestic economy			
GDP, 1987 prices[a]	4,232.0	4,407.0	4,589.0
Per capita GDP (US$[a])	258.0	264.0	267.0
Percent change	0.8	2.3	1.1
Industrial growth (%)	7.5	14.1	8.0
Agricultural growth (%)	2.9	2.5	3.2
Annual inflation rate	22.4	32.1	41.5
Balance of payments			
Exports (FOB)	178.0	170.0	150.0
Imports (CIF)	491.0	468.0	513.0
Trade balance	−313.2	−291.1	−363.0
Current account balance	−263.0	−288.0	−120.0
External aid	445.0	409.0	406.0
Foreign debt	2,638.0	2,832.0	3,000.0
Debt service ratio (%)	94.0	105.0	106.0
Gross forex reserves	85.0	96.0	148.0
Official exchange rate[b]	428.9	734.0	1,133.8

[a]Calculated using 1987 exchange rate of Shs. 60 per US$.
[b]Shillings per US$.

SOURCE: U.S. Department of State, "Current Economic Situation and Trends, 1993" (Memo).

Although there is no question that they stimulated economic growth, these measures failed to resolve problems such as the high level of debt service commitments and excessive government spending. (Table 5.2 reviews some major economic indicators for the 1990–1992 period.) Additionally, some foreign investors and donors remain concerned about Uganda's human rights record.[11] In addition to the failure of the barter trade strategy, continuing corruption by some Ugandan officials has impeded economic development. These issues undoubtedly will continue to plague Uganda's economy until well into the next century.

Agriculture

Agriculture, which is the most important sector of Uganda's economy, accounts for about 70 percent of GDP and approximately 77 percent of the country's export earnings. The agricultural sector employs some 80 percent of

Uganda's population. Small-scale farmers form the backbone of Uganda's economy. According to a government survey, the average household owns 2.8 acres of farmland and 2 acres of other land that is used for grazing and fallow. Only 16 percent of Uganda's households are landless; however, many of these are located in urban centers. In 1991–1992, the agricultural sector grew by 3.6 percent. According to the government, much of this growth occurred in cash crop production as a result of price incentives offered to farmers that reflected competitive exchange rates and the liberalization of the marketing system. The output of food crops experienced only a 0.3 percent growth rate, largely because of widespread drought conditions.[12]

Coffee is the main cash and export crop. (Table 5.3 reviews coffee exports for the 1980–1992 period.) Robusta coffee, which accounts for more than 90 percent of the country's coffee-producing areas, is cultivated around Lake Victoria and other southern regions at altitudes from 4,000 to 5,000 feet. Arabica coffee is grown in Mbale District at altitudes ranging from 5,000 to 7,500 feet. Coffee earnings have been depressed because of the July 1989 collapse of the ICO agreement, which had maintained a quota system for prices. Within twelve months, the price for 2.2 pounds of robusta fell from $2 to $0.50. Since Uganda relied on coffee for about 97 percent of its foreign exchange earnings, the ICO collapse virtually bankrupted the government. As a result, several donors, including Italy and the African Development Bank, suspended loans to Uganda because of mounting repayment arrears. Uganda averted a financial disaster when various donors pledged $640 million through the World Bank for project and balance of payment support.[13]

Apart from the ICO crisis, the Museveni regime also had to contend with growing controversy surrounding the marketing of coffee. At issue was the interest payments of crop finance. Farmers complained that the Coffee Marketing Board—which was the sole legal buyer, processor, and exporter of coffee—failed to pay them on time. When the government failed to respond to their protests, the hard-pressed farmers refused to sell their coffee. In July 1987, the Coffee Marketing Board tried to resolve this problem by buying coffee directly from the farmers, thereby avoiding cooperative unions, which traditionally had acted as intermediaries. This decision naturally alienated the cooperative unions, which wanted to protect their commissions.

Officials ended this debate by freeing the coffee industry from the grip of the inefficient Coffee Marketing Board. The Ugandan government therefore removed the board's monopoly and reorganized it as a limited liability company. At the same time, Kampala licensed four cooperative unions in Banyankole, Busoga, Bugisu, and Masaka and authorized them to make coffee shipments abroad through the facilities of the Union Export Services association.[14] This new

TABLE 5.3   Exports, 1980–1992

	Quantity (tons)	Earnings (US$ thousands)
Coffee		
1980	110,100	341,300
1981	128,300	243,800
1982	174,700	349,400
1983	144,300	346,300
1984	133,200	359,600
1985	151,500	348,500
1986	140,800	394,200
1987	148,153	307,535
1988	144,254	265,279
1989	176,453	262,811
1990	141,489	140,384
1991	127,438	120,794
1992	119,795	95,699
Cotton		
1980	2,300	4,100
1981	1,200	2,300
1982	1,800	3,200
1983	7,000	11,200
1984	6,700	12,100
1985	9,553	13,979
1986	4,875	5,086
1987	3,443	4,097
1988	2,088	2,968
1989	2,321	4,020
1990	3,808	5,795
1991	7,819	11,731
1992	7,740	8,488
Tea		
1980	500	300
1981	500	300
1982	1,200	800
1983	1,300	1,200
1984	2,500	3,300
1985	1,200	1,000
1986	2,800	3,100
1987	2,100	1,900
1988	3,079	3,079
1989	3,195	3,194
1990	4,760	3,566

(continues)

TABLE 5.3 *(continued)*

	Quantity (tons)	Earnings (US$ thousands)
1991	7,018	6,780
1992	7,816	7,711
Tobacco		
1980	300	300
1981	0	0
1982	0	0
1983	700	900
1984	700	1,500
1985	300	400
1986	0	0
1987	0	0
1988	39	58
1989	490	569
1990	2,268	2,823
1991	2,467	4,540
1992	2,544	4,800

SOURCES: Ministry of Finance and Economic Planning, *Background to the Budget, 1989–1990* (Kampala: Ministry of Finance and Economic Planning, 1989), p. 135; and World Bank, *Uganda: Growing Out of Poverty* (Washington, D.C.: The World Bank, 1993), p. 182.

arrangement enabled the cooperative unions to get higher prices for their coffee. However, these cooperatives were not free to select the most cost-effective transport to move coffee to Mombasa for shipment to overseas markets. Moreover, the value of the exporter's margin, which was fixed in November 1991, declined because of the shilling's devaluation. As a result, there was an increase in the cost of processing and exporting, a substantial part of which was incurred in foreign exchange.

In early 1991, the National Resistance Council passed a bill to establish a more efficient regulatory body known as the Uganda Coffee Development Authority.[15] Approximately one year later, the government decided to make the coffee sector more competitive by removing all price controls on coffee. In mid-1992, the government launched a four-year plan to salvage the country's coffee industry from further collapse. This plan involved enhancing productivity through better husbandry practices; rehabilitating old coffee gardens through a four-year $17.6 million coffee tree replanting campaign, improving the quality of coffee for export, and ensuring competitive pricing through changes to the marketing infrastructure.[16] In mid-August 1993, Uganda and numerous other nations established an association of coffee-producing countries to guarantee higher coffee prices.

Despite these measures, the coffee market remained depressed. In 1993 (the latest available figures), Uganda earned only an estimated $108 million from coffee exports, the lowest amount in more than a decade. Also, coffee growers maintained that Uganda still suffered from a weak marketing infrastructure and a lack of adequate transportation, which undermined producer incentives. Moreover, producers and government officials agreed that it would be some time before coffee-producing nations succeeded in forcing price increases.

This pessimism ended unexpectedly in early 1994, when coffee prices on the world market showed a dramatic improvement. Uganda's coffee industry also received a boost by uncertain crop prospects in Mexico, Colombia, and Côte d'Ivoire. As a result of these developments, officials expected Uganda's coffee crop to earn $190 million in 1994, up from $108 million the previous year.

Historically, cotton has been Uganda's second most important crop. (Table 5.3 reviews cotton exports for the 1980–1992 period.) In the 1950s, cotton contributed about 25 percent of total agricultural exports, and by the early 1970s, Uganda ranked third among African cotton producers. However, low official prices and the destruction of ginneries during the Amin years caused production to decrease from 467,000 bales in 1970 to 18,800 bales in 1981. Although there have been efforts to revitalize the cotton industry, the 1990 production level was only one-twentieth of the 1970 level.[17] In 1992, however, the government reported that farmers harvested an estimated 7,740 tons, compared with a peak of 86,000 tons in 1970. The NRM government hopes to increase cotton production by increasing prices paid to farmers, expanding cotton-growing areas, and reducing the operations of the Lint Marketing Board, which has a marketing monopoly of cotton lint and cottonseed.

To accomplish these goals, officials devised a twofold strategy. On 3 May 1993, Uganda launched a $10 million project to improve cotton crop variety. Under this scheme, which was funded by the United Nations International Fund for Agricultural Development, farmers from selected districts receive high-quality seeds for a three-year period. The Ugandan government also proposed to restructure the board so that unions and private processors can market lint and seed once production reaches 100,000 bales per year.[18]

Despite these steps, there was little improvement in this sector. In 1993, cotton exports declined by 67 percent because of marketing problems and smuggling to Kenya and Zaire. According to the Ministry of Finance and Economic Planning, Uganda's exports decreased from 7,536 tons in 1992 to 2,482 in 1993. In January 1994, the Ugandan government responded to this crisis by passing legislation to end price controls on cotton. On 7 February 1994, Kampala also asked the World Bank for $45 million in soft loans to revive the cotton industry. Government officials are convinced that these measures will increase cotton production.

Throughout the 1960s, good soil conditions and favorable climate enabled tea output to grow rapidly. (Table 5.3 reviews tea exports for the 1980–1992 period.) In 1972, tea production reached a record level of 23,376 tons. This production all but ceased after the Amin regime expelled the country's Asians, who owned many of the tea estates. Continuing instability and warfare also prompted many tea farmers to reduce or stop tea cultivation. After Amin's downfall, the authorities succeeded in persuading Mitchell Cotts to return to Uganda and to establish the Toro and Mityana Tea Company in a joint venture with the government.[19] As a result, tea production increased from 1,700 tons in 1981 to 5,600 tons in 1985. Over the next few years, levels decreased because of security problems and then gradually increased as stability returned to Uganda and as international tea prices improved. In 1990, Uganda produced 6,777 tons of tea. In 1991 it produced 8,806 tons, and in 1992 more than 9,000. Nations that import Ugandan tea include Sudan, Somalia, Pakistan, Dubai, Germany, Japan, Djibouti, Great Britain, Kenya (re-exports), Yemen, Egypt, and Saudi Arabia. According to the Ministry of Planning and Economic Development, further production increases depend on improving rural transportation by rehabilitating feeder roads in tea-growing areas.[20]

During the immediate postindependence period, tobacco was a major foreign exchange earner, ranking fourth after coffee, cotton, and tea. (Table 5.3 reviews tobacco exports for the 1980–1992 period.) Most tobacco grows in the West Nile region of northwestern Uganda. As this was a particularly violent area during the Amin regime's last years, tobacco production suffered, falling from 5,000 tons in 1972 to 100 tons in 1981. After a boost in producer prices and the implementation of a $5.5 million rehabilitation program, the tobacco crop increased to 1,900 tons in 1984. Production declined to 932 tons in 1986, but with the help of the British American Tobacco (BAT) company, it increased to 1,287 tons in 1987; 2,513 tons in 1988; 3,833 tons in 1989; 4,200 tons in 1990; 4,444 tons in 1991; 6,686 tons in 1992; and 7,203 tons in 1993.[21] Uganda exports most of its tobacco to Japan, the Netherlands, and Belgium.

In early 1994, BAT revealed plans to build a new $15 million to 17 million redrying plant in Jinja. This facility will increase production capacity from the present 5,000 to 6,000 tons annually to 12,000 tons yearly. Additionally, BAT expressed its skepticism about the Ugandan government's plans to liberalize the tobacco industry by allowing uncontrolled production and marketing. According to BAT, tobacco production would probably drop 18 percent in 1994 because of low world prices and less demand.

The sugar industry nearly collapsed during the 1970s, as production fell from a 1968 high of 152,000 tons to almost nothing. In 1981, the Madhvani and Mehta families, who had been expelled along with all other Asians, returned to Uganda and started to rebuild the two largest sugar estates at Kakira and Lugazi. However, ongoing insurgencies delayed production until 1988 at Lugazi and 1990 at Kakira. In 1989, the Kinyala Sugar Works resumed production. The rehabilitation of these three estates will give Uganda a refining capacity of at least 187,000 tons a year, which should be sufficient to satisfy local demand, estimated to be 160,000 to 200,000 tons annually. As a result, Uganda will probably continue to import some sugar for domestic consumption. Moreover, locally produced sugar will remain more expensive than imported sugar until the industry reduces production costs.[22]

Uganda's main food crops are maize, finger millet, beans, sorghum, sweet potatoes, cassava, groundnuts, wheat, peas, soya beans, potatoes, matoke (plantain), and simsim. There are also three rice-growing projects in Uganda, the largest of which is in Olwiny Swamp in the north. Until 1972, Uganda was self-sufficient in food and exported a variety of commercial and food crops. During the 1972–1978 period, however, most farmers reverted to subsistence production; a 1979–1981 drought further degraded production of food and cash crops. In the late 1980s, food crop production recovered, except for parts of northern Uganda and Karamoja, which continued to suffer from drought and insecurity. In 1991 (the latest figures available), production of food crops expanded at an overall rate of 4 percent. In 1992, drought caused crop failures and food shortages in several districts, prompting the government to call attention to the fact that the country's strategic food stocks were low.[23]

The livestock industry is an important component of the Ugandan economy. Approximately 3 percent of households depend on livestock for their livelihood. A government estimate of the 1993 livestock population includes 5.3 million head of cattle, 5.22 million goats, 0.87 million sheep, 1.2 million pigs, and 21 million poultry. The greatest threats to the well-being of the cattle industry are cattle rustling, especially in Gulu, Apac, Soroti, Kitgum, and Kumi districts, and endemic diseases such as East Coast Fever, trypanosomiasis, rabies, and foot-and-mouth disease.[24]

About 20 percent of Uganda's total area—or 16,988 square miles—is covered by lakes, rivers, and swamps. Fishing therefore is an important industry, especially in rural regions. A 1988 fisheries survey showed that 20,750 households were involved in fish catching and 28,500 families worked in ancillary activities and the service sector at the landings. During 1993, fishermen netted 276,000 tons of fish, compared to 211,000 tons in 1989 and 214,000 tons in 1988.[25]

One of the greatest threats to the well-being of the fishing industry is environmental. According to various experts, Lake Victoria, which is an important component of Uganda's fishing industry, is on the verge of an ecological disaster.[26] Many scientists maintain that the 1962 introduction of the Nile perch (*Lates nilotocus*), which experts had hoped would be a new, high-yielding protein source for Uganda, Tanzania, and Kenya, was a mistake. The proliferation of this fish, a voracious predator that can weigh more than 200 pounds, now poses the single

Cattle are a vital component of Uganda's economy, especially among people such as the Banyoro. (Photo courtesy of U.S. Committee for Refugees.)

greatest threat to Lake Victoria's delicate ecological balance. This species feeds on smaller fish, including haplochromines, which, along with other species such as catfish, eat algae at the bottom of the lake. As the Nile perch reduced the population of algae-eating fish, oxygen levels dropped to almost zero, thereby choking the life out of Lake Victoria. The proliferation of the Nile perch also caused a market glut. In 1968, for example, Nile perch accounted for only 0.5 percent of the annual catch; by 1985, this figure had risen to 59 percent. This change impoverished fishermen who made their living by harvesting other species such as prawn or catfish. Additionally, since the Nile perch cannot be processed by sun drying since it is too oily, fishermen must cut trees to smoke them. Many areas around the lake have been denuded of trees, an important link in Uganda's ecosystem.

Another major threat to Lake Victoria and to Uganda's other lakes is the water hyacinth (*Eichhornia crassipes*), a large free-floating aquatic weed that forms a mat on still or slow-flowing water. Since first appearing in the late 1980s, this weed has multiplied rapidly and covered many lakes with a dense mat that deprives fish and plankton of oxygen essential for their survival. Government efforts to encourage fishermen to remove the water hyacinth failed to produce any noticeable results.[27] Therefore, in mid-1993 the Uganda Freshwater Fisheries Research Organization announced plans to import beetles from Benin, a West

African country that successfully used them to control biologically the weed in infested areas.

Other problems confronting Uganda's aquatic environments include overfishing, deforestation, rapid human population growth, and the introduction of untreated sewage into lakes, rivers, and streams. Toxic waste runoff, especially from the Kilembe copper mine into Lake George, could cause unexpected health and economic difficulties for future generations. The Ministry of Environment Protection hopes to monitor toxic wastes, but the lack of resources will inhibit such efforts for the foreseeable future.[28]

Industry

Uganda's main industries are the processing of coffee, cotton, tea, tobacco, sugar, edible oils, and dairy products; vehicle assembly; grain milling; brewing; and the manufacturing of textiles, steel, metal products, cement, soap, shoes, matches, animal feeds, paints, and fertilizers. The political and economic instability of the 1970s and early 1980s caused a drastic fall in Uganda's industrial output. In 1980, industry was operating at only 15 percent of capacity; by 1986, estimated industrial output was little more than a third of the 1970–1972 postindependence peak levels. Despite rehabilitation efforts, by 1992 the industrial sector operated at about 30 percent below capacity and provided only 4 percent of GDP. (Table 5.4 contains the index of industrial production during the 1987–1992 period.) To improve industrial output, the NRM wants to reduce the government's role and increase the private sector involvement in this sector and to improve the efficiency of those enterprises that remain under government control.[29]

There are numerous commercially important mineral deposits in Uganda. (Table 5.5 contains production statistics for selected minerals during the 1986–1991 period.) In the early 1970s, the mining sector produced copper, tin, bismuth, wolfram, tantalite, phosphates, limestone, and beryl. It employed 8,000 people and accounted for 9 percent of the country's exports. By 1979, however, internal instability and warfare had caused nearly all mines to cease operations.[30]

After 1986, equipment and foreign exchange shortages restricted activities in this sector to small mining operations that produced limited quantities of tin ore, wolfram, bismuth, beryl, gold, and tantalite. To better exploit the country's mineral resources, the NRM devised a project called "Mineral Investment Promotion" to strengthen the Geological Survey and Mines Department and to increase training for mining engineers at Makerere University. The government maintained that this project would allow the Geological Survey and Mines Department to increase exploration and evaluation of Uganda's mineral deposits.[31]

Initially, these efforts failed to result in any significant improvements. As a result, by 1988 mining accounted for less than 1 percent of GDP. By the early 1990s,

TABLE 5.4   Index of Industrial Production, 1987–1992 (annual average)

	1987	1988	1989	1990	1991	1992
Food processing	100.0	128.0	153.9	174.9	227.4	245.6
Drinks/tobacco	100.0	139.6	143.7	155.2	176.1	155.2
Textiles/clothing	100.0	121.8	132.7	116.3	110.9	111.9
Leather/footware	100.0	62.0	54.3	75.3	60.1	79.5
Timber/paper, etc.	100.0	135.1	169.4	183.6	198.2	223.4
Chemicals/paint						
and soap	100.0	111.2	164.7	183.5	192.9	250.3
Bricks/cement	100.0	94.4	108.9	154.2	162.6	203.1
Steel/steel products	100.0	87.2	95.8	107.7	149.3	190.7
Miscellaneous	100.0	134.0	204.2	181.0	251.2	271.7
Total all items	100.0	123.8	145.2	155.5	178.2	191.2

SOURCES: Economist Intelligence Unit, *Country Profile 1993–94: Uganda* (London: The Economist Intelligence Unit, 1993), p. 21; and James Tumusiime (ed.), *Uganda 30 Years 1962–1992* (Kampala: Fountain Publishers, [1993]), p. 104.

TABLE 5.5   Production of Selected Minerals, 1986–1991

	1986	1987	1988	1989	1990	1991
Gold (grams)	149.7	–	20.5	1,700	75,230	776,000
Tin ore (ton)	43.5	9.7	63.8	45.0	31.2	72.2
Wolfram (ton)	19.1	30.2	74.9	32.2	48.3	98.3
Kaolin (ton)	400	–	–	–	–	–
Feldspar (ton)	200	–	–	–	–	–
Iron ore (ton)	–	–	11.1	–	–	–
Tantalite(ton)	7.7	–	–	5.4	2.7	–
Phosphate (ton)	–	–	–	–	25	30
Gypsum (ton)	–	–	–	–	–	43.1
Limestone (ton)	–	–	–	–	385.5	807.5

SOURCE: Ministry of Planning and Economic Development, *Background to the Budget, 1992–1993* (Kampala: Ministry of Planning and Economic Development, 1992), p. 198.

however, exploration was being carried out for petroleum, gold, iron ore, nickel, volcanic ash, and lime.

Additionally, Uganda succeeded in getting North Korean help to restart the Kilembe copper mine. On 25 June 1992, Uganda also signed a joint venture agreement for the construction of a production facility at the Kilembe mine to extract

Owen Falls Hydroelectric Station, Jinja. Since its inauguration in April 1954 by Queen Elizabeth II, this facility has been a key source of electrical power. (Photo by Thomas P. Ofcansky.)

cobalt from slag dumps with the French government–owned Bureau de Recherches Géologiques et Minières and the British firm Barclay's Metals, Ltd. The project's cost is about $45 million; production is scheduled to begin in 1995.[32] To further exploit the country's mineral wealth, the government hopes to attract additional foreign investment and technical assistance.

Traditionally, charcoal and fuel wood have accounted for 75 percent of the country's commercial energy requirements; petroleum products, 21 percent; and electricity, 3 percent. To reduce pressure on the nation's dwindling forests, the NRM has sought to develop alternative energy sources. In the late 1980s, the government therefore gave priority to expanding the country's hydroelectric power capacity. This has proved to be a difficult undertaking, as there has been virtually no investment in the maintenance or expansion of the sector for the past twenty years. One of the chief goals involved a two-phase plan to increase the installed capacity at the Owen Falls hydroelectric station at Jinja from 150 to 180 and then to 282 megawatts.[33] However, a combination of technical and financial problems, exacerbated by unsettled political conditions, slowed progress on the project. Finally, on 22 January 1994, President Museveni inaugurated a Chinese- and Norwegian-supported $200 million dam and hydroelectric station project to rehabilitate and to expand output to 200 megawatts, which is far short of the original plan.

Throughout the postindependence period, Uganda has imported all its petroleum products. Since the NRM seized power, this arrangement has become increasingly burdensome as the demand for oil has increased. By the early 1990s, the country was spending more than $7 million per month on petroleum products. To lessen dependence on foreign suppliers and to satisfy domestic crude oil requirements, the NRM has authorized explorations throughout the country, especially around Lakes Albert and Edward along the Western Rift Valley. Areas identified for test drillings included Masindi, Hoima, Bundibugyo, and Kabarole; the government also reserved test blocks in Kigezi District and portions of Arua and Nebbi districts. By the early 1990s, there were seven international oil companies operating in Uganda (Shell, Total, Petrofina, Caltex, Upet, Esso, and Agip), three of which (Shell, Total, and Agip) were 50 percent owned by the Ugandan government. However, spokespersons for these companies have indicated that Uganda is just a "borderline case" for oil development.[34]

Transport and Communications

By 1986, Uganda's transportation and communications infrastructure was all but destroyed as a result of decades of instability, violence, and warfare. The NRM realized that the successful rehabilitation of the nation's economy depended on rebuilding this sector. Therefore, the government's RDP devoted 29 percent of investment to improving the road and rail system, which would facilitate the movement of agricultural and industrial products to markets throughout the country. Less attention has been given to Uganda Airlines, which remains plagued by a variety of financial, technical, and personnel problems. On a more positive note, there has been a slow, but steady, rehabilitation and expansion of Uganda's communications network.

Since 1986, road repair has been one of the NRM's highest priorities. The enormity of this undertaking should not be underestimated. Uganda has about 16,762 miles of roads, of which approximately 3,726 miles are all-weather, including 1,117 miles bituminized. At the time President Museveni assumed office, the condition of nearly all these roads was "very poor."[35] Nevertheless, there has been considerable improvement in the country's road network over the past several years.

In August 1987, for example, Uganda launched a $32.6 million Third Highway Project to recondition existing surfaced and unsurfaced roads. There also was a "northern corridor" project to connect eastern Zaire, Rwanda, and Uganda to the Kenyan port of Mombasa. In February 1987, Kampala signed an agreement with a Yugoslavian company to build a 155-mile road in western Uganda, from Mityana to Fort Portal, as part of the Trans-African Highway. Germany, Japan, the United Nations Development Program, and Banque Arabe de Développement Économique en Afrique (BADEA) have also funded a $55 million venture to rehabilitate rural roads while the European Development Fund financed repairs to Kampala's roads. By 1993, repair work continued on many of Uganda's roads. However, the major problem confronting the NRM was to ensure that regular maintenance work was done on rehabilitated roads.[36]

During the colonial period, the British built Uganda's 871-mile railway link to the Kenyan port of Mombasa, which, in turn, provided access to the world economy. Following the dissolution of East African Railways in 1977, the Uganda Railways Corporation (URC) assumed responsibility for operating and maintaining the country's rail system. This proved to be an impossible task, as rebel activity had destroyed several lines and had prevented repair and maintenance work on numerous others. As a result, by the late 1980s many sections of the railway required relaying, regrading, or realigning.

Shortly after coming to power, the NRM announced a five-year $150 million project to rehabilitate the railway, which is an important component of the "northern corridor." Great Britain, France, Italy, Germany, and the European Community (EC) assisted in this endeavor. Additionally, Uganda sought to reconstruct the 198-mile line from Kampala to the western town of Kasese, in one of the country's richest agricultural areas. In 1989, Italy extended a $65 million loan to Uganda to repair about 62 miles of track and in July 1991 pledged another $58 million for the rehabilitation of the line. However, continued maintenance problems and derailments eventually forced the government to condemn the Kampala-Kasese line.[37] Most experts believed it would be some time before rail service to Kasese could operate safely and efficiently.

Despite this setback, there has been some progress toward restoring the country's rail system. On 7 January 1992, President Museveni announced the opening of the Kampala–Port Bell line and a wagon ferry terminal at the port, which gives Uganda access to an alternative sea outlet in Tanzania. The previous year, President Museveni confirmed that plans were under way to build a railway line from Musoma (northern Tanzania) to the port of Tanga (northeastern Tanzania), which Uganda could use as an alternative to the port of Mombasa, Kenya. Ending Uganda's reliance on Mombasa is one of Museveni's long-standing goals. Apart from the high costs of using Mombasa, Museveni believes that the often tenuous political relations between the two countries could result in a closure of Mombasa to Ugandan goods. However, financial and technical problems will prevent the construction of a Musoma-Tanga line anytime soon.[38]

In July 1993, the URC resumed services to northwestern Uganda after an eight-year suspension caused by instability in the region. Initially, there was weekly passenger and cargo service along the 174-mile route from Gulu to Pakwach. The Ugandan government promised to increase service as demand grew. Additionally, Kampala unveiled long-term plans to extend the northern railway line to Oraba on the Ugandan-Sudanese border.

The URC also increased the size of its fleet to 73 locomotives, 101 passenger coaches, and 1,699 goods wagons and built a modern locomotive workshop at Nalukolongo, near Kampala. Moreover, the URC acquired two mobile 40 ton–capacity container-handling cranes. These improvements enabled the volume of traffic to increase from 254,000 tons in 1985 to more than 500,000 tons in 1991.[39]

Notwithstanding these accomplishments, the URC remains one of Uganda's four big loss-making corporations. (The other three are Uganda Airlines Corporation [UAC], Uganda Transport Company [UTC], and Uganda Posts and Telecommunications [UP&TC].) Moreover, the URC faced numerous mainte- nance and financial problems. For example, shortly after commencing operations in 1992, one of the three wagon ships operating from Port Bell was down for re- pairs. Also, the Kampala-Kasese line lost at least Shs. 24 million a month. Financial problems forced the layoff of 819 employees. Most important, interna- tional donors such as the World Bank and Germany, which has threatened to cut its aid, have expressed dissatisfaction with the URC's performance.[40]

Civil aviation has had a troubled history in Uganda. In 1931, the Imperial Airways started a weekly seaplane flight from London to Port Bell and Butiaba. Four years later, the colonial authorities opened an international airport at Entebbe, which helped expand Uganda's links to the outside world. On 1 January 1946, the East African Airways Corporation inaugurated air service to various African, European, Middle Eastern, and Asian destinations. Several other airlines, including British Overseas Airways Corporation and Central African Airways, also operated to Uganda.

Immediately following independence, civil aviation in Uganda prospered. Until the early 1970s, more than twenty scheduled international carriers served Entebbe, but many of them terminated their operations during Idi Amin's brutal and chaotic regime. Additionally, the East African Airways Corporation collapsed when Uganda, Kenya, and Tanzania disestablished the EAC. In May 1976, the gov- ernment established UAC; however, it was not until February 1978 that the airline inaugurated its first scheduled international service, between Entebbe and Nairobi. In the early 1980s, UAC started passenger and cargo service to Europe. However, chronic internal instability and warfare hampered the airline's opera- tions.

As a result, by 1986 UAC was suffering from numerous financial, technical, and personnel problems. However, the greatest difficulty confronting the airline was the condition of its aging fleet. In 1988, the British government banned UAC Boeing 707 flights to London because of excessive noise levels. Officials arranged to have a "hushkit" (a device to reduce engine noise) used on one of UAC's Boeing 707s so that flights to Britain could continue; this aircraft operated only a short time before crashing in Rome. By the late 1980s, UAC operated two Fokker F-27s and an old "unhushkitted" Boeing 707. From January 1989 to February 1990, the airline operated a DC-10 leased from Ghana Airways to and from London. Financial problems caused UAC to cancel this arrangement.[41]

As UAC continued to sink further into debt, the Ugandan government decided to seek help from a British-based consulting firm, PSAIR, to restructure and reor- ganize the airline. In mid-1989, this company released a report that was highly critical of UAC operations. Among other things, PSAIR claimed that the airline had a bad reputation and a poor safety record. Moreover, PSAIR said that UAC's customer service and reliability were substandard and suggested that the airline close its catering facilities at Entebbe International Airport. PSAIR also noted that UAC lacked financial controls on its operations and, as a result, had sunk into insolvency. The PSAIR report recommended that UAC employ four expatriates to resolve its difficulties. Although it accepted the report, the government failed to act quickly on PSAIR's recommendations. Indeed, it was not until 24 July 1991 that the minister of works, transport and communications, Ruhakana Rugunda, announced that UAC's staff would be reduced from 629 to 328, with the help of PSAIR.[42] As a result, airline service continued to deteriorate.

On 6 February 1990, conditions worsened when UAC pilots and engineers went on strike for higher wages. The Ministry of Works, Transport, and Communications responded to UAC's various ills by halting its operations for three months. During this time, the airline cut all routes except that to and from London and reduced its staff to 100.[43] On 8 April 1990, UAC resumed domestic flights from Entebbe to Arua and Kasese and regional service between Entebbe and Nairobi. However, as nothing was done to resolve the airline's many problems, the UAC again degenerated into chaos.

By mid-1991, the airline owed its creditors more than $9.4 million and operated only one small aircraft, which made six weekly flights to Nairobi and one internal flight to Arua. To increase service, Uganda signed an agreement with Tanzania and Zambia to form African Joint Air Service, which would operate flights from Africa to Europe and India.[44] Financial difficulties prevented the three countries from starting African Joint Air Service operations.

The following year, continuing financial and labor problems, coupled with the inability to acquire additional aircraft, pushed UAC to the verge of bankruptcy. The airline therefore sold off part of its assets to improve its financial position and to meet more than $400,000 in unemployment benefits to laid-off staff. UAC further reduced costs by closing all its overseas offices except in Nairobi and Dubai.[45]

By 1993, UAC had improved its financial condition, largely because of a World Bank–funded reorganization and restructuring program. Additionally, the airline acquired a new $10.6 million Boeing 737-200, which enabled it to initiate twice-weekly passenger and once-weekly cargo flights to Johannesburg, South Africa, via Harare, Zimbabwe. On 22 May 1993, Uganda and Swaziland signed an agreement to commence passenger and cargo flights between the two countries in 1994. As long as it continues to receive Western aid, UAC should be able to maintain this modest regional service.

Prior to the EAC's collapse in 1977, the East African Posts and Telecommunications Corporation maintained Uganda's post and telecommunications systems. After the destruction of the 1970s and early 1980s, the UP&TC started to rebuild the country's communications infrastructure. By the late 1980s, UP&TC furnished relatively good service for those who lived in cities such as Kampala and Entebbe. Long-distance communications are via a radio-relay system in Kampala.

There is also a 960-channel radio-relay link between Kampala and Nairobi. A satellite ground station at Mpoma operates two antennae, one with the International Telecommunications Satellite Corporation (Intelsat) Atlantic Ocean satellite and the other with the Intelsat Indian Ocean satellite. By late 1990, there were 57,185 telephones in use nationwide. On 8 August 1994, the Ugandan government announced its decision to privatize postal and telecommunications services to improve efficiency.[46] Despite this step, most industry observers believed that it would be some time before there was significant improvement in these sectors.

Recent UP&TC activities included laying underground telecommunication cables in Kampala, Jinja, and Entebbe and completing High Frequency Radiocall installations in twenty-two stations throughout the country. Additionally, the UP&TC has agreed to install modern digital electronic exchanges at various district headquarters. There will be thirty-five channels for Kampala-Kasese, ten channels for Kampala-Mubende, and ten channels for Kampala-Mityana. The UP&TC also has reopened and upgraded numerous sub–post offices. Domestic Expedited Mail Service links have grown from twenty-three to forty-five offices, and international Expedited Mail Service links have expanded from twenty-one to forty countries.[47]

Tourism

From the 1960s to 1972, Uganda was renowned as one of East Africa's most important tourist destinations. At its peak in 1971, when the number of tourist arrivals totaled 85,000, tourism was Uganda's third foreign exchange earner, after coffee and cotton, and was a major source of direct and indirect employment. The tourist industry rested on three national parks, thirteen game reserves, and eight animal sanctuaries.[48] Within these havens, there was a staggering array of fauna, including elephant, lion, white and black rhinoceros, giraffe, buffalo, waterbuck, hippopotamus, topi, Uganda kob, bushbuck, mountain gorilla, chimpanzee, eland, zebra, leopard, and crocodile. Additionally, there were hundreds of migratory bird species throughout Uganda.

During Idi Amin's regime, which lasted from 1971 until 1979, the tourism sector suffered terribly from neglect and organized poaching. As a result, some species such as the white rhinoceros disappeared completely while most other animals were drastically reduced in number. Additionally, widespread fighting destroyed most tourist hotels, roads, and other facilities. Under Milton Obote's second presidency from 1980 until 1985, officials planned to revive the country's tourist industry with Western assistance and to restore wildlife herds by implementing a ban on hunting. Although conditions improved somewhat and the number of tourists slowly increased, Uganda remained a remote destination compared to Kenya and Tanzania, its most popular neighbors.

After seizing power in 1986, Yoweri Museveni embarked on a major rehabilitation program to boost tourism. Apart from rebuilding tourist hotels and lodges, the government promoted conservation and wildlife management. The improvement in security and the return of international airlines such as British Airways, Sabena, and Egypt Air to Uganda also encouraged tourism. However, despite these encouraging developments, many problems still plague the tourism sector, the most important of which are human encroachment and poaching in areas reserved for wildlife. Financial difficulties also hamper the operations of the Uganda National Parks (UNP) and the Game Department, the two institutions responsible for managing and protecting the country's wildlife.

Since 1988, the UNDP has funded a $1.56 million "Support to Wildlife and National Park Management" project, which seeks to rehabilitate wildlife resources in Uganda and strengthen UNP and the Game Department. Additionally, the EC has financed a "Conservation of National Resources" program, which, apart from various training efforts, seeks to preserve natural resources in the Murchison Falls, Queen Elizabeth, and Kidepo Valley national parks and in Lakes Victoria, Edward, George, and Albert and Kidepo Basin. Some of the other public and private international agencies involved in conservation work in Uganda include USAID, the World Wide Fund for Nature (formerly known as the World Wildlife Fund), and Care and Wildlife Conservation International.[49]

The Ugandan government has also taken steps to preserve the country's natural heritage. In August 1991, UNP director Eric Edroma announced that arrangements had been finalized to establish a wildlife college in Queen Elizabeth National Park to train Ugandans in wildlife management. Additionally, the NRM has expanded Uganda's national park system to include the Bwindi Impenetrable Forest Reserve, which is one of the last refuges of the mountain gorilla; the Elgon Forest Reserve; and the Ruwenzori Forest Reserve. There are also eleven wildlife sanctuaries.[50]

These and numerous other endeavors have scored some successes. In early 1991, for example, UNP reported that the elephant population in the national parks had doubled from about 1,000 to more than 3,000 over the past decade.[51] However, such accomplishments are the exception rather than the rule. According to the International Union of Conservation of Nature and Natural Resources (IUCN), Uganda's square-lipped rhinoceros, black rhinoceros, and cheetah are extinct and the mountain gorilla, chimpanzee, Ruwenzori blackfronted ducker, and golden monkey are on the brink of extinction. Numerous other species, including the elephant and the Nile crocodile, are on the IUCN's red (most endangered) list of threatened animals.

Apart from the absence or paucity of wildlife and continuing instability in or around many national parks, the other threat to Uganda's fledgling tourist industry concerns the inadequate amenities available to foreign visitors. Many hotels are substandard and lack supplies, proper equipment, and trained staff. This problem is particularly acute in the national parks, wildlife sanctuaries, and conservation areas. The difficulty of traveling to and especially within Uganda also hampers tourism.[52] In 1993, when the number of tourist arrivals totaled an estimated 82,000, the government unveiled a $75 million program to rehabilitate hotels and game lodges to lure tourists back to Uganda. The following year, the NRM launched a campaign to attract $100 million in foreign investment to improve the tourist industry.

In mid-1993, the Ugandan government started an improvement program in the national parks by authorizing five tour-operating firms to undertake several construction and rehabilitation projects. According to the agreement, Abercombie and Kent will build a hotel and lodging facilities in Bindwe National Park. In Murchison Falls National Park, Sarova Hotels International of Kenya will rehabilitate, expand, and manage Paraa Safari Lodge while the Kilimanjaro Safari Club will renovate and expand Pakuba Grand Lodge. Uganda Holidays will undertake a similar project at Chobe Lodge. Lastly, Ruwenzori Mountaineering Service will improve facilities in Ruwenzori National Park.[53]

On 29 November 1993, Uganda's Minister of Tourism, Wildlife, and Antiquities, James Wapakabulo, unveiled a ten-year plan to further improve the tourist industry. As part of this plan, the ministry abolished visa requirements for people visiting Uganda from countries such as the United States, Canada, Japan, and Australia; simplified entry and exit procedures; and permitted direct charter flights from neighboring countries to tourist sites without stopping at Entebbe International Airport. The government also has plans to establish twenty-three tourist centers around the country to help revive the tourism sector.[54] Despite the government's commitment to rehabilitating the tourist industry, other priorities and the increasing difficulty of attracting foreign economic assistance mean it will probably be some years before Uganda becomes a major East African tourist destination.

Uganda in the World Economy

Like most Third World nations, Uganda depends on Western financial assistance, technology, goods, and services for its economic well-being. South-south and regional economic cooperation—though politically attractive and in some cases economically beneficial—contribute very little to the varied needs of a country like Uganda, which is trying to rebuild an infrastructure destroyed by almost two decades of warfare. By the early 1990s, it became clear that continued access to Western aid would be dependent on Uganda's respect for human rights; progress toward democratization; and willingness to adopt seemingly harsh, curative economic policies formulated by institutions like the IMF. Pursuing such policies has been difficult for a government that has yet to eliminate all armed opposition and to establish its economic credibility in the world marketplace. Nevertheless, the

NRM has no choice but to rely on the West for the resources required to stimulate Uganda's economic development. At the same time, Kampala will continue maintaining economic relations with nations throughout East Africa, especially Kenya, which is its chief trading partner, and the Third World.

Historically, agricultural products, mainly coffee, tea, and cotton, have dominated Uganda's exports. During the early 1980s, exports increased, enabling Uganda to post a $65.7 million surplus in 1984, a $115 million surplus in 1985, and a $2.2 million surplus in 1986. During the Museveni period, the country's main trading partners have included Kenya, Great Britain, the United States, the Netherlands, Germany, and France. Other important trading partners have been Japan, India, Belgium, Spain, Italy, and the members of the former Preferential Trade Area for Eastern and Southern Africa (PTA). (Tables 5.6 and 5.7 reflect the export and import levels of Uganda's main trading partners.)

Shortly after the NRM seized power, the country's shrinking foreign exchange reserves, coupled with an overvalued shilling, caused a drastic reduction in imports. In May 1987, the NRM responded to this crisis by implementing a 76 percent devaluation of the shilling. As the foreign exchange crisis deepened, Uganda concluded several barter trade agreements with many nations and companies. However, when it became evident that this strategy had failed, the government unveiled a scheme whereby properly licensed private companies retained foreign exchange earned for nontraditional exports such as fruits and vegetables. These companies then could sell any or all of their foreign exchange to the Central Bank and within 180 days apply for licenses valued at the equivalent of their foreign exchange earnings to finance imports. Additionally, Uganda established a $12.5 million export trade promotion credit with USAID to facilitate the production and marketing of several nontraditional exports, including seeds, fertilizers, jute, tin, and packaging materials.

To resolve the balance of trade dilemma, the NRM enacted new import-export licensing procedures. Banks processed imports designated as "foreign exchange required." For imports not involving foreign exchange payment, the government required an import license. All exports needed a Ministry of Trade license, which listed the amount of foreign currency involved in the transaction and declared receipts to the Central Bank. By the early 1990s, the NRM had started to encourage private incentive by moving away from the practice of requiring government agencies such as the Coffee Marketing Board to process all exports.

In January 1992, the government further liberalized foreign exchange allocations by introducing an auction system, whereby importers can purchase foreign exchange in a timely manner. This decision completed the transformation from a fixed exchange rate to one determined by market forces and reduced the excessive controls associated with an administered exchange rate system. Inevitably, the shilling's value declined, from a level of 400 per U.S. dollar to 700 per U.S. dollar. This devaluation in turn reduced import duty on a range of goods. Some of the

TABLE 5.6   Value of Exports, Main Trading Partners (percent of total value)

	U.S.	UK	France	Spain	Netherlands
1986	29.4	15.7	9.0	11.0	6.1
1987	25.0	17.8	11.3	9.7	9.1
1988	17.2	15.1	11.7	9.3	11.9
1989	14.1	11.2	11.8	9.7	14.5
1990	8.0	10.7	13.2	4.5	22.1
1991	10.8	6.3	14.7	10.1	19.6

SOURCES: Ministry of Finance and Economic Planning, *Background to the Budget, 1992–1993* (Kampala: Ministry of Finance and Economic Planning, 1992), p. 157; and Economist Intelligence Unit, *Country Profile 1993–94: Uganda* (London: Economist Intelligence Unit, 1993), p. 28.

TABLE 5.7   Value of Imports, Main Trading Partners (percent of total value)

	Italy	UK	Kenya	Japan	Germany
1986	5.3	11.4	31.6	4.8	10.6
1987	12.6	13.4	17.8	7.0	9.4
1988	7.3	14.3	21.3	5.8	7.1
1989	4.9	14.9	19.6	6.5	11.2
1990	10.5	16.0	20.2	5.6	10.0
1991	6.0	15.0	23.3	9.3	6.9

SOURCES: Ministry of Finance and Economic Planning, *Background to the Budget, 1992–1993* (Kampala: Ministry of Finance and Economic Planning, 1992), p. 157; and Economist Intelligence Unit, *Country Profile 1993–94: Uganda* (London: Economist Intelligence Unit, 1993), p. 28.

more common items included a reduction from 50 to 30 percent on textiles, clothing, footwear, and headgear; 20 to 10 percent for tires for bicycles and motor vehicles; and 50 to 20 percent for lubricating oil.

In April 1992, officials sought to protect the coffee industry's profitability in the face of falling world prices by allowing coffee exporters to surrender foreign exchange proceeds to the Bank of Uganda at the average exchange rate given by the forex bureau. This eliminated an implicit tax that had placed coffee at a disadvantage with regard to other exports.[55]

Throughout the NRM era, Uganda's balance of payments have suffered from increasingly high deficits. In 1987, the outstanding external debt was about $1.4 billion, which represented more than half of GDP and almost three and a half times the level of exports. By 1989, the external debt had risen to $1.8 billion, and by 1990 it had climbed to almost $2 billion, the servicing of which accounted for 54.5 percent of exports. In 1991, the external debt reached $2 billion, or 106 percent of exports. During his 1992–1993 budget speech, Joshua Mayanja Nkangi announced that he had decided to create a Central Debt Unit in the Ministry of Finance and Economic Planning to manage Uganda's growing external debt.[56]

Since 1986, Uganda has tried to resolve the growing balance of payments problem by negotiating a revised repayment program. In June 1987, for example, the Paris Club rescheduled $113 million in loans and in 1989 rescheduled another $93.1 million. In June 1992, the Paris Club recommended that Uganda's external debt, which by then had reached about $2.6 billion, be waived by 50 percent to alleviate balance of payments and debt-serving burdens. This plan will be followed by a rescheduling of the remaining debt at market rates over twenty-three years, with a six-year grace period.[57]

Rebuilding Uganda's political, economic, and social infrastructures required substantial amounts of foreign economic assistance. Since 1986, Uganda has acquired at least $2 billion from numerous Western donors to finance an array of reconstruction projects.[58] (Table 5.8 contains bilateral and multilateral aid levels for the 1986–1991 period.) This relationship started shortly after Yoweri Museveni assumed the presidency and proposed a six-month $160 million emergency relief and rehabilitation program. Before pledging any funds, the international donor community waited until the NRM established an economic policy. In early 1987, the NRM, the World Bank, and the IMF conducted discussions about Uganda's reconstruction. After the Ugandan government devised the RDP, the IMF agreed to fund a three-year package that included a one-year $24 million Structural Adjustment Facility and $32 million over the next two years. Additionally, the IMF provided $20 million from its Compensatory Financing Facility fund. The World Bank financed a $100 million program to support the RDP. A June 1987 World Bank Consultative Group meeting resulted in further commitments totaling $310 million for the first year of the RDP. At about the same time, the Paris Club rescheduled $66 million in debts. In February 1988, the IMF approved a $33.7 million compensatory financing facility in connection with Uganda's shortfall in export earnings for the year ending September 1987, caused primarily by low world coffee prices. In October 1988, a World Bank Consultative Group meeting of international donors committed $550 million to Uganda in concessionary loans and grants for 1989.

When the NRM launched its ERP, the World Bank released the $50 million second tranche of its Structural Adjustment Facility, after Uganda repaid $18 million of arrears to the IMF. In April 1989, the IMF sanctioned a three-year low-interest Enhanced Structural Adjustment Facility worth $179.3 million in Special Drawing Rights. The World Bank also approved a $25 million supplement to the first Economic Recovery Credit. In 1990 it approved a $265 million supplement, $125 million of which represented balance of payments support, which the bank had withheld in 1989 pending Uganda's implementation of anti-inflation measures. In late 1989, international donors pledged $640 million. On 7 May 1992,

TABLE 5.8   Gross Official Development Aid (millions of U.S. dollars)

Bilateral	U.S.	UK	Sweden	Germany
1986	4.0	10.8	1.9	12.9
1987	14.0	12.7	4.0	15.8
1988	18.0	49.9	11.2	18.7
1989	21.0	40.7	16.3	19.9
1990	39.0	36.0	14.5	27.0
1991	51.0	51.5	34.3	34.2
Multilateral	IDA	IMF	EC	ADF
1986	61.7	–	16.8	0.2
1987	111.3	–	31.8	0.9
1988	68.0	40.2	40.0	19.1
1989	92.0	54.2	36.4	21.4
1990	193.0	81.1	36.0	23.2
1991	137.0	78.4	30.1	24.3

Total
1986   208.3 of which grants 131.5
1987   287.3 of which grants 153.5
1988   422.4 of which grants 262.7
1989   506.2 of which grants 237.8
1990   655.7 of which grants 311.1
1991   626.4 of which grants 371.2

SOURCES: Economist Intelligence Unit, *Country Profile 1993–94: Uganda* (London: Economist Intelligence Unit, 1993), p. 30; and Ministry of Planning and Economic Development, *Background to the Budget, 1991–1992* (Kampala: Ministry of Planning and Economic Development, 1991), p. 177.

the World Bank announced a $71 million soft loan to finance reconstruction of northern and northeastern Uganda. (The loan covers the districts of Gulu, Kitgum, Lira, Apac, Moyo, Nebbi, Arua, Soroti, Kumi, Tororo, and Pallisa.) According to the loan's terms, Uganda would finance 10 percent of the estimated three-year $100 million program, which covers the areas of agriculture, water and sanitation, urban and rural feeder roads, highways, education and training, urban and community development, and telecommunications. Then, on 18–19 May 1992, a World Bank Consultative Group meeting of Western donors pledged $850 million for Uganda's 1992–1993 structural reform program.[59]

The NRM is committed to playing a major economic role in regional affairs. Apart from participating in normal trade activities, Uganda belongs to four regional economic organizations. These organizations include the PTA, which has been transformed into the Common Market for Eastern and Southern Africa (COMESA); the Lomé Convention; the Kagera Basin Organization (KBO); and the Inter-Governmental Authority on Drought and Development (IGADD). Additionally, the Ugandan government has supported the reestablishment of the EAC.

The PTA comprised twenty-one member states from east, central, and southern Africa: Uganda, Angola, Burundi, Comoros, Djibouti, Eritrea, Ethiopia, Kenya, Lesotho, Madagascar, Malawi, Mauritius, Mozambique, Namibia, Rwanda, Seychelles, Somalia, Swaziland, Tanzania, Zambia, and Zimbabwe. It advocated the creation of a common market; free trade; and cooperation in industry, agriculture, transport, and communications. On 20 January 1993, Yoweri Museveni was elected PTA chairman at the organization's tenth summit in Lusaka, Zambia. He promised to help devise a program to ensure food security in the region and to facilitate intra-PTA economic cooperation through the removal of trade barriers and import licence restrictions. Historically, the PTA's performance had been limited by the so-called rules of origin, which authorized preferential treatment only for goods manufactured by companies in which citizens of the member state in question manage the enterprise and hold at least 51 percent equity in the company.

On 5 November 1993, at the third PTA summit, Uganda and fourteen other nations (Eritrea, Ethiopia, Kenya, Lesotho, Madagascar, Malawi, Mauritius, Mozambique, Namibia, Rwanda, Sudan, Swaziland, Tanzania, and Zambia) signed a treaty that transformed the PTA into a common market known as the Common Market for Eastern and Southern Africa (COMESA).[60] This new organization, which is more ambitious than its predecessor, will seek to enhance regional cooperation in numerous areas, including trade, customs, monetary cooperation, transport and communications, industry, energy, environment, agriculture, and legal, administrative, and budgetary matters. COMESA plans to establish a fully functioning common market by the year 2000 and to create a monetary union by the year 2020.[61] The treaty also included a provision for COMESA to impose sanctions on any member state that refuses to comply with the organization's decisions.

The Lomé Convention is a trade and aid agreement between the EC and sixty-six African, Caribbean, and Pacific nations. The convention guarantees duty-free entry to the EC for designated commodities from these countries. As a member state, Uganda has not only benefited from this agreement but has also received aid from the European Development Fund, which provides assistance to all countries that have signed the Lomé Convention.[62]

In 1977, Tanzania, Rwanda, and Burundi established the KBO; in 1981, Uganda joined the organization. The KBO promotes the economic development of the 23,166–square mile Kagera River basin. The organization has numerous ambitious plans, including the construction of a hydroelectric dam at Rusumo, Rwanda; a railway to link all four countries; and various road construction schemes throughout the region.[63] However, all KBO projects lack funding, because the member states have failed to meet their financial obligations to the organization. At a December 1991 meeting between the member states, Rwanda, Tanzania, and Burundi promised to pay their arrears; the delegates also agreed to apply a commercial interest rate to future unpaid contributions.

Despite its financial problems, the KBO has scored some successes. In 1990, for example, the organization launched a three-year tsetse fly control project that, if successful, would increase livestock population by 10 percent.[64] On 22 September 1992, the KBO inaugurated direct telephone links between Uganda, Tanzania, Burundi, and Rwanda without detouring lines through Europe as before.

In January 1986, Uganda, Djibouti, Ethiopia, Kenya, Somalia, and Sudan (Eritrea joined the organization after achieving its independence in 1993) created the IGADD to deal with the problems of drought, desertification, and agricultural development. Like the KBO, IGADD suffers from a shortage of money, largely because its members do not pay their annual contributions. As a result, IGADD has accomplished little of economic benefit. However, the organization has provided the member states a forum to resolve some of their political problems.

The first two IGADD summits helped Ethiopia and Somalia end their dispute of the Ogaden region. At the 1990 summit, the six nations pledged to promote the peaceful resolution of the region's many internal conflicts by supporting each other's peace and reconciliation efforts. On 15 May 1992, IGADD joined the newly created Observatoire du Sahara et du Sahel, which seeks to increase cooperation among its twenty member states in the battle against drought and desert encroachment.[65] On 6 September 1993, the IGADD opened its fourth summit in Addis Ababa, Ethiopia. Apart from calling attention to the economic and social ramifications of population migration and recurrent drought, famine, and desertification, the member states failed to devise any strategies to stimulate economic growth throughout the region.

One positive result of the summit was that IGADD agreed to participate in the stalled peace process in southern Sudan. However, most observers agreed that the IGADD peace initiative was unlikely to achieve any results as long as the Sudanese government and the factions of the insurgent Sudanese People's Liberation Army (SPLA) remained committed to a military solution of the political problems. Moreover, the organization appeared to be increasingly moribund. A five-year (1992 through 1996) program designed to enhance food security, improve communications, and protect the environment has accomplished little because the member states have failed to adequately finance IGADD. For example, out of an approved $1.3 million budget for 1992, the IGADD secretariat received only $383,186 in contributions from Djibouti and Ethiopia. By the end of 1992, the rest of the member states were in arrears totaling almost $2 million. The IGADD executive secretary, David S. Muduuli, who is from Uganda, downplayed the organization's poor financial condition by saying that IGADD's progress should not be based on money.

The Museveni regime has also supported efforts to reestablish the EAC. In October 1992, for example, Ugandan, Kenyan, and Tanzanian officials met in Arusha, Tanzania, and worked out the details for a protocol on cooperation in trade, industry, agriculture, health, finance, transport, and research and communications. On 30 November 1993, the presidents of these three countries ratified this protocol by signing an East African cooperation treaty. This document committed Uganda, Kenya, and Tanzania "to strive for greater political and economic cooperation among themselves and within the wider concept of African unity and economic integration." According to President Museveni, a revived EAC would stimulate economic growth, facilitate communication within East Africa, encourage the adoption of common tax policies, and help Uganda to manage its external debt.[66]

6

INDEPENDENCE AND NEW FOREIGN POLICY DIRECTIONS

Sɪɴᴄᴇ ɪᴛ ɪs ʟᴀɴᴅʟᴏᴄᴋᴇᴅ, Uganda depends on foreign imports for consumer goods, luxury items, and energy requirements. The country's economic well-being also depends on receiving regular amounts of foreign financial aid. Additionally, all postindependence governments sought varying amounts of foreign military assistance, largely for use against the Ugandan population or numerous indigenous rebel groups. As a result of these factors, Uganda's foreign policy is predicated upon maintaining good relations with those nations capable of fulfilling its economic and military needs.

After independence, the Western countries that performed this role included Great Britain, Israel, and the United States. However, like many African governments, Kampala sought closer relations with radical states such as the People's Republic of China (PRC), the former Soviet Union, Libya, and North Korea. By accepting help from any nation, Uganda hoped to avoid becoming a pawn in the cold war and to preserve a nonaligned reputation, especially as far as the United States and the former Soviet Union were concerned.

Ugandan foreign policy since 1962 can be divided into three phases. During Milton Obote's first presidency, which lasted from 1962 to 1971, the Ugandan government concentrated primarily on acquiring financial aid to facilitate social and economic development. During the 1971–1986 period, Uganda's foreign policy reflected an increasing need for military assistance as the country degenerated into warfare and anarchy. In the last phase, which began when Yoweri Museveni seized power in 1986, Uganda started placing more emphasis on its need for economic rather than military aid.

Uganda in the World Arena

During the immediate postindependence period, Anglo-Ugandan relations remained close.[1] London hoped to preserve its special political and economic links with Uganda. The British government also wanted to persuade Kampala to establish a democratic government, pursue a pro-Western foreign policy, and institute a free market economy. Uganda looked to Britain for economic and military assistance. As a result, by the end of 1967 Britain was the main foreign investor in Uganda (48.1 percent of all foreign investment). In the military sphere, several British military personnel served in Uganda, and a British officer acted as commander of the Ugandan army. British troops also helped the Ugandan government suppress a 1964 mutiny by Ugandan troops.

Despite these mutual interests, there were disagreements between the two countries. Britain was against Milton Obote's "Move to the Left" and his opposition to British arms sales to South Africa. As a result, London welcomed Obote's demise and became the first country to recognize Idi Amin's government, which promised to maintain Uganda's historic economic ties to Britain. The British also provided military assistance to Amin, despite growing evidence of his complicity in human rights violations. However, after Amin expelled Uganda's Asian community in 1972, Britain terminated its aid program and imposed an economic embargo against Uganda. In January 1973, Amin responded to these developments by recalling Uganda's high commissioner from London, nationalizing British tea estates and several other firms, and threatening to expel 7,000 British residents from Uganda. Although it severed diplomatic relations with Uganda in July 1976, the British government allowed private companies to furnish luxury goods to the Amin regime in exchange for Ugandan coffee.

After Amin's downfall, Britain restored diplomatic relations with Uganda and promised to provide the country's new government with economic and technical assistance. However, within a few months, Godfrey Binaisa ousted Yusuf Lule's regime. This created a dilemma for the British government, which thought that Binaisa might have been laying the groundwork for Milton Obote's return. Thus, London postponed the recognition issue by claiming that it was "not clear whether the need arises for recognition."[2] Despite this political hesitation, Britain provided training, vehicles, and uniforms to Uganda's police force and about £4 million in economic aid. As soon as Binaisa's regime collapsed, the British canceled the £4 million assistance program but agreed to finance one-third of the cost of the 1980 election, which returned Obote to power.

Despite its initial caution, Britain eventually provided considerable support to Obote after he convinced Whitehall that he had abandoned his socialist ideology in favor of a mixed economy. Apart from deploying a thirty-man Special Air Service training team to Uganda, the British government supported decisions by several British companies that had previously operated in the country to return and restart their former businesses. Additionally, British engineers rebuilt Kampala's water system, and numerous British banks and accounting firms helped Uganda to reestablish its banking industry and revive its economy. London also canceled £22 million in debt and allocated £3.5 million for new development programs.[3]

Even after evidence surfaced that implicated Obote in massive human rights violations, Britain continued to provide assistance to Uganda. To rebuild and train the Ugandan police, the British government furnished sixty-three Land Rovers, £800,000 to rehabilitate police barracks, and security training for Ugandan officers in Britain. The British also contributed personnel to a thirty-six-person Commonwealth military training team, which also included troops from Canada, Australia, Jamaica, Guyana, Sierra Leone, Kenya, and Tanzania. When the Commonwealth states withdrew from the team in March 1984 to protest Uganda's poor human rights record, Britain continued to train Ugandan security personnel. Whitehall justified its policy by claiming that the Soviet Union and North Korea would step into any vacuum created as a result of a British departure from Uganda. Despite increasingly international criticism, Britain supported the Obote regime until its downfall.

After seizing power in January 1986, Yoweri Museveni expressed a desire to maintain cordial ties with the British. However, relations between the two countries became strained over the British Military Advisory Training Team's activities. According to the minister of state for defence, Ronald Batta, the British were "arrogant" toward Ugandans and traveled around the country without permission. As a result, the team departed Uganda in mid-November 1986 and Kampala arranged for Tanzanian military personnel to help train the NRA.[4] Museveni also criticized the Thatcher government for its refusal to impose economic sanctions against South Africa.

Despite these difficulties, Anglo-Ugandan relations steadily improved over the next several years. On 8 March 1990, the British minister of overseas development, Lynda Chalker, announced that the British government had decided to increase its aid to Uganda from £10 million to £15 million during the 1990–1991 period.[5] Most of this and subsequent British assistance helped to rehabilitate the Owen Falls Dam and the Uganda Electricity Board and to improve the country's judiciary, police, and Ministry of Planning and Economic Development. Additionally, Britain helped to train the Ugandan police. In late May 1994, the British government purchased 75.4 tons of food for famine victims in Soroti and Kumi districts. Two months later, Chalker informed President Museveni that the British government had pledged £20 million to Uganda for the 1994–1995 fiscal year. Of this sum, £3 million would be spent in activities connected with the NRA's demobilization program.

In 1962, Israel established diplomatic relations with Uganda, largely because Tel Aviv saw a unique opportunity to isolate and destabilize Sudan, one of its potential Arab League enemies. Since its independence in 1956, Sudan, which considers itself an Arab nation, has been divided between the Muslim, Arab north and the Christian/animist black African south. These differences caused southern guerrillas known as the Anya Nya to revolt against the north. From Israel's point of view, as long as this war continued, Sudan would be too busy to fight Israel. Therefore, Tel Aviv used Uganda as a conduit to supply arms and training to the Anya Nya guerrillas.

In exchange, Uganda received military, paramilitary, intelligence, and police training for its security personnel. In August 1963, four Ugandans also qualified as pilots on a Piper Super Cub in Israel. By 1965, Tel Aviv had provided Uganda with small arms, light artillery pieces, and other military equipment. Additionally, Israel helped establish the Ugandan Air Force and equipped it with Piper Super Cub and Piaggio aircraft. After Congolese (Zairian) planes bombed western Ugandan villages in 1965, Israel provided Uganda with six armed Fouga Magister jet trainers and three C-47 Dakota transport planes. By early 1967, Israel had reassigned approximately fifty instructors to staff training schools for Ugandan pilots, artillery officers, and paratroopers. In nonmilitary areas, Tel Aviv furnished Uganda with technical cooperation, agricultural assistance, community and social work courses, and engineering scholarships and organized a national youth movement.[6]

Shortly after seizing power in 1971, Idi Amin made two visits to Israel with requests for military and economic aid. When Tel Aviv refused to supply him with the kind of military hardware he wanted, relations between the two countries quickly deteriorated. In late February 1971, Amin accused the Israelis of building secret bases in Uganda. The following month, he purged twenty Israeli-trained pilots from the Ugandan Air Force. On 30 March 1971, the Ugandan government closed Israel's embassy and ordered all Israelis to leave the country.[7]

For the next several years, Amin exploited every opportunity to criticize Israel and to praise Tel Aviv's enemies. Tension between the two countries reached a crescendo on 28 June 1976, when Amin allowed a hijacked Air France airbus with 250 hostages to land at Entebbe International Airport. The Popular Front for the Liberation of Palestine (PFLP) claimed responsibility for this act of piracy and demanded the release of 53 Palestinian prisoners in Israel and several other nations. The hijackers then freed all the passengers except for 100 Israeli citizens and the crew that stayed with them and threatened to kill them if their demands were not met. Prior to the expiration of the PFLP's final ultimatum, a force of Israeli commandos raided Entebbe on 3 July 1976 and freed the hostages.[8]

After Amin severed diplomatic relations in 1973 following the Arab-Israeli war, Kampala and Tel Aviv had no official dealings with one another for more than two decades. However, in late 1992, Uganda lifted immigration bans on Israelis. Then, in May 1993, Ugandan vice-president and minister of internal affairs, Samson Kisekka, visited Israel. Shortly thereafter, a Ugandan military delegation also made a trip to Israel. By late 1993, there were a small number of Israeli experts in Uganda working on various agriculture projects. Finally, on 29 July 1994, the two nations resumed diplomatic relations. Kampala and Tel Aviv hope to expand commercial, educational, and cultural ties.

The United States was one of the first Western nations to recognize Uganda's independence. Over the next three decades, Washington tended to pursue a policy that stressed respect for human rights and economic cooperation. Political relations between the two countries have often been strained, especially during the Amin years and Milton Obote's second presidency. After 1986 the United States gradually assumed a more active role in Uganda, and by the early 1990s it had become one of the country's most important Western allies.

During the immediate postindependence period, the United States purchased more than 20 percent of Uganda's exports. Despite the importance of this trade, Uganda frequently found itself at odds with the United States. In late 1964, for example, the Uganda People's Congress (UPC) Youth League issued a statement that condemned U.S. participation in the Stanleyville raid in the Congo (now Zaire).[9] The Ugandan press and parliament then debated the political costs of accepting U.S. aid and whether Washington had adopted a neocolonial policy in Africa.[10]

Shortly after Amin seized power, Ugandan-U.S. relations deteriorated. After Amin expelled Uganda's Asians, Washington cut its aid to Kampala. In 1972 the United States evacuated 112 Peace Corps personnel from Uganda, and the following year it closed its embassy in Kampala. Between 1974 and 1976, the United States did little to increase pressure against Amin. In early 1977, Washington started to reevaluate its policy after the International Commission of Jurists reported that the number of dead in Uganda had risen to about 400,000. In October 1978, the U.S. Congress, under pressure from various human rights organizations, passed a bill that imposed a total trade ban on Uganda.[11]

After Amin's downfall, the United States resumed diplomatic relations and lifted its trade embargo. However, Washington was slow to resume its aid to Uganda, and by 1982 its activities were limited to a $4 million grant to assist cooperatives and to rehabilitate agriculture. Growing dissatisfaction with Milton Obote's human rights record eventually caused a rupture between the two countries. In August 1984, the U.S. assistant secretary for human rights, Elliott Abrams, issued a report that described Uganda's human rights situation as being among the worst in the world. Obote responded by ordering the withdrawal of the U.S. military attaché and threatening to suspend the U.S.-Uganda aid agreement.

Although it promised to invest $100 million in Uganda between 1985 and 1988, Washington maintained its distance from the increasingly brutal Obote regime.[12]

After 1986, there was an immediate improvement in U.S.-Ugandan relations. Washington pledged $10 million annually for basic reconstruction and a loan for rehabilitation of the agricultural sector and restoration of coffee plantations and tanneries. Despite its opposition to Uganda's relations with Libya and other radical states, the United States continued to provide economic assistance to Uganda.[13] In 1986, for example, USAID launched a $4.3 million West Nile agricultural rehabilitation project. By late 1988, Washington had earmarked grants and loans totaling $29.8 million. The following year, U.S. officials announced plans to implement an International Military Education and Training program that in 1990 and 1991 brought Ugandans to the United States for command and staff training, infantry officer courses, medical training, and courses in vehicle maintenance.[14] In 1991, the Peace Corps resumed operations in Uganda after a seventeen-year absence; twelve volunteers worked in the areas of education and resource management. On 26 July 1994, USAID granted $8 million to Uganda to finance its primary education reform program. Apart from this aid, the United States has encouraged the Museveni regime to make political reforms, promote democratization, and downsize the NRA.

Despite this cooperation, the two nations have had disagreements, particularly about Uganda's human rights record and Kampala's supposed connection to the insurgent Rwanda Patriotic Front (RPF). Another problem surfaced on 18 August 1992, when U.S. customs agents detained and questioned Stephen Kapimpina Katenta-Apuli, Uganda's ambassador to the United States, about his role in an illegal scheme to purchase 400 antitank missiles.[15] The American authorities eventually dropped all charges against the ambassador and released him. This incident, although embarrassing to both countries, failed to have a lasting impact on Uganda-U.S. relations.

Indeed, over the next two years, relations between Kampala and Washington grew much closer. Economically, Uganda benefited from ever increasing amounts of U.S. aid. In late 1993, for example, USAID provided a $9.3 million grant to finance a program to improve family planning services and maternal health and signed a six-year $25 million agreement to increase and diversify nontraditional agricultural exports. The following year, the two nations concluded an $8 million agreement that supported the Uganda Primary Education Reform Program, which sought to improve teacher salaries and conditions of service and to provide grants to schools nationwide. In June 1994, President Museveni visited the United States to seek trade and investment from firms and individuals in Nebraska, Texas, and Washington, D.C. According to the Ugandan government, he received considerable private and public assistance for his regime.

In the political arena, Uganda and the United States cooperated on a range of issues, including the IGADD-sponsored peace talks for southern Sudan and peacekeeping efforts in Somalia and Liberia. On 24 July 1994, the Ugandan government also welcomed the arrival of U.S. military forces in Kampala, which signaled the beginning of "Operation Sustain Hope." This massive relief effort provided food, medicine, clean water, and other supplies to hundreds of thousands of Rwandan refugees who had fled their country to camps in eastern Zaire in the wake of the genocide against Rwanda's Tutsi population carried out by members of the Habyarimana regime and Hutu extremists.

Given its need for financial and technical aid, Uganda will seek to remain a close U.S. ally. For its part, Washington will continue to rely on Kampala, which wields considerable influence throughout East Africa, for support of its policies. Given these mutual interests, most observers agree that Uganda and the United States will enjoy a productive relationship for the foreseeable future.

Ugandan-Chinese relations originated because Beijing wished to block Tel Aviv's efforts to gain a pro-Western foothold in East Africa. To neutralize Israeli influence, the PRC supplied a range of economic and military assistance to Kampala. In 1965, the Chinese provided Uganda with about a $3 million grant and $12.4 million in credits.[16] Additionally, Beijing sent some small arms and a military aid mission to Uganda. However, in late 1967, after Ugandan officers complained that the Chinese mission had engaged in revolutionary activity and had distributed lapel buttons displaying Mao Zedong's picture, Prime Minister Obote expelled the mission.

Relations between the two countries remained strained during the early months of Idi Amin's regime. Throughout 1971 Amin repeatedly accused the PRC of supporting Tanzanian and Obote's efforts to destabilize his regime. In April 1972, however, Kampala and Beijing resolved their differences, largely through the efforts of Amin and the Chinese ambassador to Uganda. Over the next few years, Uganda and the PRC issued numerous statements praising one another's government. Amin also sent a military attaché to Beijing, and claimed that the PRC had offered to sell arms to Uganda. During the Ugandan-Tanzanian war, Beijing blamed "Soviet social-imperialism [for] doing its utmost to aggravate divergences, create contradictions and even repeatedly provoke bloody conflicts among African countries."[17]

Despite these seemingly close ties, the PRC welcomed the end of Amin's regime, largely because of its military ties to Moscow. However, it was not until the early 1980s that Beijing provided any assistance to Uganda. The two countries signed agreements to study the feasibility of rice growing and to improve Ugandan fisheries. The PRC also granted about $210,000 to Uganda to purchase spare parts for the Kibimba Rice Company. In early 1982, a Chinese military mission visited Uganda, the results of which were never made public.[18]

After Yoweri Museveni seized power in 1986, there was further improvement in Ugandan-Chinese relations. Apart from an array of economic aid, the PRC and Uganda concluded several joint ventures to establish bicycle, paper, construction, and manufacturing factories. Additionally, Uganda signed a $73 million contract with the Chinese construction company Sietco for construction of a second hydroelectric power station at Jinja. Chinese medical teams also performed two-year tours in Jinja Hospital. Additionally, Beijing agreed to finance the construction of Nelson Mandela Freedom Stadium at Namboole, east of Kampala.[19]

In contrast to China's relatively minor role in Uganda, the former Soviet Union was one of Kampala's closest allies. In July 1965, the two countries signed a military agreement whereby the former Soviet Union supplied a squadron of two MiG-15 and four MiG-17F fighter-interceptors, airfield ground support, military maintenance facilities, ground-to-ground and ground-to-air radio communication equipment, artillery pieces, and military trucks. Although this matériel was free of charge, Uganda had to pay for spare parts and ammunition. Moscow also trained more than 250 Ugandan army personnel, 20 pilots, and 50 air force technicians and mechanics. By late 1967, 25 Soviet advisers were in Uganda helping to integrate this equipment into the Ugandan armed forces.

During the 1970s, the former Soviet Union, hoping to increase its influence in East Africa, increased its military assistance to Uganda. In July 1972, a Ugandan military delegation visited Moscow and arranged to take delivery of several weapon systems, including tanks, armored personnel carriers, missiles, transport aircraft, helicopters, marine patrol boats, field engineering equipment, MiG-21s, and radar. During the 1974–1975 period, Uganda obtained more than US$500 million in military equipment. Significant items included 12 MiG-21s, 8 MiG-17s, 60 T-34/T-54 tanks, 100 armored personnel carriers, 50 antiaircraft guns, 200 antitank missiles, 850 bombs and rockets, 9 radar units, 2 Mi-8 helicopters, 250 surface-to-air missiles, 6 patrol boats, 6 mobile bridges, an unknown number of trucks and jeeps, and quantities of ammunition, spare parts, and test equipment. In addition, between 1973 and 1975 more than 700 Ugandan military personnel received training in the former Soviet Union, and more than 100 Soviet instructors managed many training programs in Uganda.

Ugandan-Soviet relations cooled in 1975, when Amin expelled the Soviet ambassador because of a disagreement over Moscow's intervention in Angola. After being embarrassed and threatened by the Israeli raid on Entebbe International Airport, Amin improved relations with the former Soviet Union. Moscow resumed arms shipments and signed a series of technical and cultural protocols with Kampala, but relations soured once again as the Amin regime deteriorated in the late 1970s.

After Amin's downfall, the former Soviet Union sought to establish friendly relations with his successors. However, the legacy of Soviet military aid to Amin undercut these efforts. As a result, Ugandan-Soviet relations remained cool for many years. Nevertheless, Moscow furnished Uganda with modest amounts of nonmilitary assistance. In 1979, for example, the former Soviet Union financed the rehabilitation of the Uganda Spinning Mill at Lira. Two years later, Kampala and

Moscow signed a Scientific and Cultural Cooperation Agreement, which provided ninety scholarships for Ugandan students and fourteen lecturers to teach scientific and technical courses at Makerere University. In August 1983, the Soviet government agreed to reschedule for fifteen years the repayment of a $37 million debt incurred during the Amin period.

Ugandan-Soviet relations gradually improved under Yoweri Museveni's presidency. Between 1986 and 1988, the former Soviet Union provided more than US$20 million in weapons to Uganda. In November 1988, the Ugandan Ministry of Defense began talks with a Soviet aircraft manufacturer to purchase an An-32 transport plane; however, Uganda never took delivery of the aircraft. By mid-1989, Moscow had halted military aid to Uganda as part of its commitment to reduce its military role in sub-Saharan Africa.

Economically, Moscow supplied the Museveni regime with student scholarships, technicians to help establish the *New Vision* newspaper, and medical equipment to Mulago Hospital. In June 1987, the former Soviet Union opened a Uganda-Soviet Cultural Friendship Society branch in Hoima. In 1988, the two nations embarked on a variety of joint projects, primarily in the agricultural sector.[20]

After the Soviet Union's collapse, the Russian Republic announced that it had assumed responsibility for all obligations of the former regime. The Russian Republic also indicated its intention to continue economic relations with Museveni's government by establishing processing plants for coffee, tea, cocoa, meat, milk, furniture, window glass, and glassware in Uganda. In addition, because of hard currency shortages in both countries, Moscow urged Kampala to increase trade links such as barter, compensation deals, and trade in small quantities. However, the level of Uganda's trade with Russia is small compared with Western nations such as the United States and Great Britain. Despite the expressions of goodwill between Kampala and Moscow, severe political and economic problems in Russia will undoubtedly prevent Moscow from significantly increasing its activities in Uganda for the foreseeable future.

Ugandan-Libyan relations began in 1972, after Amin ordered the expulsion of all Israelis from the country. Amin's anti-Israeli stance attracted the friendship and support of Libyan leader Muammar Qadhafi. As a result, over the next several years Libya provided Uganda with military and economic assistance. When Tanzania invaded Uganda in 1979, a decline in Soviet military assistance forced Amin to look to Libya for additional support. Tripoli, hoping to save the Amin regime, responded by sending an array of arms to Uganda, including three BM-21 "Stalin organ" rocket launchers and a Soviet-built Tu-22 bomber. In addition, Qadhafi deployed about 2,000 ill-trained members of the Libyan militia to Uganda.[21] Additionally, he supported several hundred Palestine Liberation Organization guerrillas who also participated in the unsuccessful fight to save the Amin regime, which collapsed in April 1979.

Like the Soviet Union, Libya's post-1979 relations with Uganda suffered because of its association with Amin. During Milton Obote's second presidency, the gulf between the two countries widened after Ugandan authorities discovered that Qadhafi was providing support to various antigovernment guerrilla groups. In September 1981, Brigadier Moses Ali, leader of the Uganda National Rescue Front (UNRF); Yoweri Museveni, commander of the NRA; and Andrew Kayiira, chief of the Uganda Freedom Movement, visited Tripoli and concluded an alliance against Obote. Although it supplied all these groups with military aid, Libya eventually threw its weight behind Museveni. Throughout his four years in the bush, Museveni downplayed or denied his links to Libya. By pursuing this subterfuge, Museveni hoped to convince Ugandans that the NRA was independent of external support. He also did not want to alienate the West by publicizing the NRA ties to Libya.[22]

After the NRA seized power, Libya furnished considerable military and economic assistance to Uganda. By early 1988, Tripoli, which had become Uganda's main arms supplier, had provided the NRA with an array of weapons, including aircraft, antiaircraft artillery, multiple rocket launchers, and small arms and ammunition. Beginning in late 1988, Libyan military aid started to decline. On 1 June 1992, Libya recalled its twenty-two maintenance technicians and engineers who had been assigned to the Ugandan Air Force.[23] Although neither Tripoli nor Kampala offered a reason for the termination of this program, many Western observers believed that many of the Libyan personnel had failed to keep the air force flying and had treated their Ugandan counterparts with contempt.

In the economic arena, Uganda and Libya cooperated on numerous projects. In April 1987, for example, the two countries concluded a $60 million barter trade deal under which they agreed to exchange Ugandan agricultural produce for Libyan oil, tractors, and cement. On 29 December 1988, the Joint Permanent Ministerial Commission of Co-operation convened its first meeting and discussed matters pertaining to trade, transport, energy, finance, investment, education, vocational training, culture, and information. Approximately one month later, Uganda and Libya concluded three economic agreements. These agreements included a $130 million Libyan loan to Uganda with a 4 percent interest rate and a twelve-year repayment period, rescheduling of the $11 million loan granted by Libya to Uganda in 1974, and a general agreement to strengthen economic cooperation. Since then, the two nations have signed numerous barter deals and economic accords. On 9 August 1994, for example, Uganda and Libya concluded an agreement to enhance cooperation in the areas of agriculture, trade, and technology. Additionally, Kampala and Tripoli reaffirmed their commitment to exchanging business information and enhancing tourism.[24]

Ugandan–North Korean relations started in December 1981, when Milton Obote visited North Korea and signed a cooperation agreement that covered numerous technical, scientific, economic, and cultural subjects. Pyongyang also agreed to deploy a thirty-man military team to Uganda, primarily to manage maintenance projects and infantry training in Gulu. Over the next few years, the North Korean military contingent in Uganda grew to about 150. Apart from their training activities, the North Koreans often led UNLA combat units in the field against antigovernment guerrillas. Such operations reportedly claimed the lives of at least three North Koreans. In September 1985, a few months after Obote had been deposed, the North Korean contingent left Uganda.

In 1986, Yoweri Museveni invited the North Koreans to return to Uganda. He wanted them to train NRA personnel in the use of North Korean equipment. Since then, North Korean advisers have helped train Ugandan military, police, and security personnel. Additionally, Pyongyang has supplied a variety of military assistance. In late 1987, for example, a North Korean consignment of weapons off-loaded at Dar es Salaam for transshipment to Uganda. This delivery included Soviet-built SA-7 surface-to-air missiles, antiaircraft guns, truck-mounted rocket launchers, armored personnel carriers, and ammunition.[25] Similar shipments of military equipment continued into the early 1990s.

The two countries also fostered economic cooperation through barter trade deals, which usually involved an exchange of Ugandan agricultural produce for North Korean finished products such as tractors. In mid-1990, Pyongyang offered to rehabilitate the Kilembe copper mine, develop an irrigation system in western Uganda, and build small-scale hydropower stations in northeastern Uganda. On 17 June 1993, the two countries signed a three-year agreement to exchange economic information. According to President Museveni, other possible areas of Ugandan–North Korean cooperation include road building and agricultural mechanization.[26]

Uganda's Regional Relations

After gaining independence in 1962, Uganda enjoyed a brief period of tranquility with its East African neighbors, especially Kenya and Tanzania. In 1967, for example, the three nations created the East African Community (EAC), which sought to create a common market and share the cost of transport, banking, postal, and communications facilities. By the late 1960s, Uganda's relations with Kenya and Tanzania became strained as Milton Obote's government moved toward economic socialism and political radicalism. After Idi Amin seized power in 1971, relations between the three countries became more acrimonious as the escalating warfare and brutality in Uganda threatened the entire region's stability. After Amin's downfall in 1979, Kampala slowly distanced itself from Dar es Salaam, which had invaded Uganda to overthrow the Amin regime. Ties with Kenya, however, gradually improved as the two countries enjoyed a prosperous economic relationship. Relations between Kampala and Nairobi grew tense after the emergence of the Museveni regime

Instability, warfare, political radicalism, border unrest, and a variety of refugee problems complicated Uganda's relations with the Sudan, Rwanda, and Zaire. Despite these difficulties, there was considerable cross-border petty trading, smuggling, and arms trafficking between Uganda and all three nations. Additionally, Uganda carried on legitimate commercial activity with each of these countries.

The 1962–1986 Period

A civil war in southern Sudan largely determined the state of relations between Kampala and Khartoum. Between 1962 and 1969, Uganda sympathized with black African southern Sudanese rebels who fought the Arab-dominated northern government for greater autonomy. By the end of this period, approximately 74,000 refugees from war-torn southern Sudan had settled in northern Uganda. After a group of young, radical Sudanese military officers—led by Colonel Jaafar Nimeiri—seized power in 1969, Ugandan president Milton Obote switched his loyalties to the Sudanese government to reaffirm his own radical credentials. After the 1972 Addis Ababa Agreement ended the civil war in southern Sudan, many southern Sudanese joined the Ugandan army. After Amin's downfall, these soldiers, along with tens of thousands of Amin's Ugandan supporters, fled to southern Sudan. Another 200,000 Ugandan refugees fled to southern Sudan during Obote's second presidency. In 1983, a second civil war started in southern Sudan, which caused yet another flow of refugees into Uganda.

From 1962 to 1986, Ugandan-Zairian relations experienced several periods of considerable instability. After 1964, political unrest in Zaire forced thousands of refugees to flee to Uganda. A military crisis between the two countries developed after Zairian premier Moise Tshombe used European mercenaries against what he considered to be disloyal rebels. Uganda responded by deploying a battalion of troops to Zaire to fight pro-Tshombe forces. On 13 February 1965, Zairian warplanes bombed two border villages, Paidha and Goli, in the West Nile District. Eventually, relations between the two countries improved after the less confrontational Mobutu Sese Seko seized power in late 1965 and Uganda severed ties to Zairian rebel groups.

During the Amin years, a rift again emerged between Kampala and Kinshasa largely because of personal rivalry between Amin and Mobutu. Also, Uganda believed that Zaire was conspiring with Tanzania to support guerrilla attacks on Uganda by anti-Amin groups. After Amin visited Zaire in 1975, relations between the two leaders gradually improved.

Ugandan refugees en route from southern Sudan to a refugee center at Yumbe, Uganda. Given the chronic instability along the Ugandan-Sudanese border, refugees will continue to depend on help available at places such as Yumbe. (Photo by Frederick Ehrenreich.)

In the early 1980s, Kampala and Kinshasa repeatedly pledged to cooperate to end banditry, smuggling, and other illegal activities along their common border. However, little was done by either country to stabilize the lawlessness, largely because both lacked the resources to exert control over their respective borders.

Historically, the nature and scope of Ugandan-Rwandan relations have been determined largely by the repercussions from clashes between Tutsi and Hutu ethnic groups in Rwanda. Between 1961 and 1966, Tutsi exiles in Uganda, Burundi, Zaire, and Tanzania launched at least ten attacks on the Hutu-dominated Rwandan government. The fighting, which failed to bring about a change in the government, caused approximately 68,000 Rwandan refugees to flee to Uganda. Kampala allowed these refugees to settle with their cattle in camps in the Ankole region. In 1966, Kigali enacted legislation that barred the return of these refugees to Rwanda. In 1969, President Milton Obote ordered the removal of all nonskilled foreigners from public employment, including thousands of Rwandans.

After Idi Amin seized power, the plight of the Rwandan refugees eased somewhat as an unknown number of them joined the Ugandan army and the security services. Additionally, many Rwandans moved out of their refugee camps onto Ankole lands. This caused tension with the Bahima of Ankole, who complained that the Rwandans and their cattle had taken over their grazing lands.

During Milton Obote's second presidency, the status of the Rwandan refugees deteriorated as the government accused them of being Amin supporters. Obote also suspected the Rwandans of providing aid to the NRM, which opposed his government. In October 1982, government-inspired violence against the Rwandans compelled about 40,000 of them to return to Rwanda. Under international pressure, Uganda and Rwanda concluded an agreement whereby Kampala accepted responsibility for all Rwandan refugees in the country. Such persecution persuaded many Rwandans to join the NRA, which was fighting to overthrow the Obote regime. The 1982 explosions also convinced many Rwandan refugees that Uganda would never fully accept them. This radicalized the group that ultimately became the core of the RPF. By the mid-1980s, Rwandans probably constituted the third-largest group in the NRA after the Banankole and Baganda.[27]

Regional Policy Under the National Resistance Movement

When Museveni seized power in January 1986, he promised to pursue a foreign policy that would enhance regional stability and cooperation. Since then, however, Uganda's relations with its neighbors, with the exception of Tanzania, have become increasingly strained and acrimonious. Armed clashes have occurred between Uganda and Zaire, the Sudan, Kenya, and Rwanda. Several factors caused these confrontations. All the countries in the area lack the resources to control the respective borders and to prevent cross-border raids, smuggling, and arms trafficking. More important, many East African nations, including Uganda, harbor and frequently provide support to one another's insurgent groups. As a result of these problems, much of the region suffers from perpetual instability.

Tanzania is Uganda's most important regional ally. Understanding this relationship requires an appreciation of Tanzania's role in Uganda after the overthrow of the Amin regime. To help Kampala restore peace, the Tanzania People's Defence Force (TPDF) maintained about 20,000 troops in Uganda. Additionally, Tanzanian soldiers managed a large-scale army training program at Mbarara. However, by mid-1980 tension between TPDF personnel and many Ugandan soldiers prompted Dar es Salaam to withdraw about one-half of its troops. Despite this decision and continuing clashes between Tanzanian troops and Ugandan citizens, President Nyerere deployed a 1,000-person police unit to Uganda. By mid-1982, Nyerere had decided to reduce and then eliminate Tanzania's military presence in Uganda, citing the high cost of maintaining troops in a foreign country.

Ugandan-Tanzanian relations improved after Museveni came to power. Militarily, Tanzania initiated a security assistance program. In late 1986, about thirty military advisers replaced a British Military Advisory Training Team that had left Uganda. Although the British returned to Uganda in January 1987 with a small, nonresident training team, Tanzanians continued to serve as advisers and trainers. During the 1988–1992 period, Tanzanian instructors managed portions of Uganda's basic training program. In addition, many NRA troops studied at Tanzania's National Military Leadership Academy in Monduli, Tanzania, and the School of Infantry at Nachingwea.[28] On 10 October 1992, the two nations signed a military training exchange agreement to facilitate peacetime cooperation between the NRA and the TPDF.

The two nations also cooperated on several economic projects. Throughout the late 1980s and early 1990s, Kampala and Dar es Salaam expressed the need to work together in the areas of aviation, research, energy, coffee production, investment, and communications. Uganda also promised to move about 60 percent of all its imports and exports through the port of Dar es Salaam. However, the difficulty of moving such large amounts of goods overland through Tanzania prevented Uganda from achieving this goal. An even more ambitious scheme involved the proposed construction of a 621 mile–long rail line from Tanga to Kampala via Arusha and Musoma.[29] Although they lack the technology to build such a railway, Uganda and Tanzania remain committed to this project and, in mid-1993, unsuccessfully sought Chinese assistance for conducting a survey of the proposed route. Although the lack of resources has delayed or hampered many Ugandan-Tanzanian development projects, the two nations remain committed to maintaining close political and economic relations.

Historically, Ugandan-Zairian relations have been complicated by the presence of refugees and rebels on both sides of the border and by smuggling, which has precipitated security problems. In May 1986, Museveni tried to resolve these dilemmas by closing Uganda's border with Zaire. Shortly thereafter, the two countries agreed to cooperate on security matters along the common border, implement antismuggling measures, and create a common market at Bwera and Akasenda for the local inhabitants.[30]

On 31 May 1987, Kampala sought to further mollify Kinshasa by ordering the NRA to attack a Zairian rebel camp on the Ugandan-Zairian border in the Ruwenzori Mountains. The Ugandan troops killed five guerrillas and captured five others, all of whom belonged to the Mouvement National Congolais (MNC).[31] Despite this action, the Zairian government continued to believe that the NRM was providing support to the MNC and another anti-Mobutu rebel group known as the Parti de Libération Congolaise (PLC). As a result, over the next few months there were several clashes between Zaire's Forces Armées Zairoises (FAZ) and the NRA on Uganda's northwestern border. Also, Kinshasa retaliated against the NRM by providing assistance to a secessionist Ruwenzori insurgent movement that had been fighting successive Ugandan governments since independence and by directing the Zairian air force to bomb anti-Mobutu rebel camps in Uganda. The two countries tried to resolve their disputes by signing a communiqué on 20 August 1987 in Kasese that required both governments to strengthen border security. The NRA therefore deployed thousands of troops along the border to stop Zairian rebels from entering Uganda.[32]

As a result of these actions, tensions between Kampala and Kinshasa eased. In June 1988, however, Ugandan-Zairian relations deteriorated when the PLC, presumably operating from bases in Uganda, mounted a series of attacks in northeastern Zaire. The PLC claimed to have killed 120 FAZ troops and wounded many others. When the FAZ launched a counteroffensive, a PLC commander led his forces into Ugandan territory. The NRA arrested nineteen PLC soldiers, incarcerated them at the local military barracks, and complained that Zaire had failed to control its dissidents. In late 1988, FAZ troops conducted several counterinsurgency operations at various locations in northwestern Uganda. When Museveni protested this incursion, Zaire closed its border with Uganda.

Relations between the two countries became more strained when former Ugandan president Idi Amin appeared in Zaire in January 1989. Holding a false Zairian passport, Amin arrived in Kinshasa aboard a regularly scheduled Air Zaire flight from Libreville, Congo. He apparently intended to return to Uganda with an estimated 500 armed supporters who were to meet him in northeastern Zaire. Museveni requested the former president's extradition, intending to try Amin for atrocities committed during his eight-year reign. Kinshasa rejected this request because there was no extradition treaty between Uganda and Zaire. Instead, the Mobutu regime detained Amin in Kinshasa and expelled him from the country nine days later. Thereafter, relations between Kampala and Kinshasa cooled, leading to the mutual expulsion of ambassadors. On 8 September 1989 the two countries restored diplomatic ties. Reports also surfaced of fighting between NRA forces and Zairian-assisted pro-Amin rebels in eastern Zaire.

Throughout 1990, the two countries worked to stabilize their common border. On 10 April, the fourth meeting of the Uganda-Zaire permanent joint commission ended in Kinshasa. Among other things, the two sides agreed to create a subcommission to deal with judicial, security, and defense matters. Later that same month, there was a border meeting in Bunia, where Ugandan and Zairian officials agreed to apprehend and repatriate runaway criminals. In July, Ugandan and Zairian officials met in Kasese and held talks on a variety of security, trade, poaching, and smuggling problems. The failure of these meetings to achieve any progress prompted FAZ units to temporarily seal off the Zairian-Uganda border at Butogota and Ishasha on 4 October in an effort to stop fugitives from crossing into Uganda.

In 1991, similar difficulties prevented the two nations from establishing harmonious relations. One issue involved an unknown number of Zairian families that had been living unlawfully in Bwindi Forest for more than four years. Some of these Zairians had started farming whereas others had earned a livelihood by illegal timber harvesting. Ugandan officials promised to repatriate these families to Zaire. A more serious problem occurred on 30 September and again on 29 October, when FAZ troops went on looting sprees in Kasindi. Although Kampala protested these incidents, there was little Kinshasa could do to stabilize its border with Uganda. Indeed, political unrest on the Zairian side of the border forced many Zairians to seek refuge in Uganda. As a result, on 4 December the Zairian government closed all entry posts into Uganda. Despite this action, border problems continued to plague the two nations. By March 1992, for example, about 20,000 Zairians had fled into Bundibugyo, Uganda, to escape harassment by FAZ troops.[33]

Kampala rejected Kinshasa's demands to repatriate these refugees because "it was contrary to the United Nations charter." Additionally, the Ugandan government authorized several nongovernmental organizations, including the International Committee of the Red Cross, OXFAM, Lutheran World Federation, World Food Programme, and the League of Red Cross and Red Crescent Societies, to provide food and other assistance to these refugees.[34]

During the 1993–1994 period, Zairian refugees continued to flee into western Uganda to escape tribal violence while bandits and self-styled rebels persisted in launching cross-border raids into western Uganda. On 7 January 1993, Governor Jean-Pierre Kamumbo Mbokko of North Kivu Province unsuccessfully proposed a joint military operation with Uganda to restore peace to the border region. As a result, conditions along the border remained unsettled. Additionally, Zairian troops periodically attacked PLC units in western Uganda. In 1993, for example, fighting between FAZ troops and various rebel organizations and clashes among many ethnic groups caused at least 16,000 Zairians to flee their homes and seek refuge in camps in Uganda's southwestern district of Kasese. In June 1993, Uganda and Zaire tried to stabilize the border area by establishing a joint commission and by investigating the possibility of military cooperation in unstable border regions. These efforts failed to produce any lasting results and, in early 1994, another wave of Zairian refugees, which numbered more than 10,000, crossed into western Uganda and settled in villages near Bundibugyo.

Given the remoteness of the Ugandan-Zairian border and the fact that both countries place a low priority on resolving their differences, it is likely that chronic instability along the border will continue for the foreseeable future. As far as Uganda is concerned, this situation could become a political problem for President Museveni, especially among western groups that already believe that Kampala has ignored their needs and concerns.

Ugandan-Sudanese relations have been characterized largely by problems associated with rebels and refugees. Historically, Ugandan insurgents have sought safe haven and assistance from supporters who lived in southern Sudan. Similarly, rebels who belonged to the Sudanese People's Liberation Army often looked to northern Uganda as a source of supplies or as a place to escape Sudanese government forces. Uganda and the Sudan have also had to cope with large numbers of refugees from one another's territory. After Idi Amin's downfall in 1979, for example, several hundred thousand Ugandans from northwestern Uganda, the deposed dictator's birthplace, fled to the Sudan to escape possible retribution by the UNLA.[35] Since then, these refugees, many of whom had supported Amin, have been a source of tension between Kampala and Khartoum.

Museveni's first problem with the Sudan emerged immediately after the downfall of Tito Lutwa Okello's regime. Remnants from the UNLA took refuge in several garrison towns in southern Sudan's Equatoria region. Khartoum used UNLA personnel to defend these garrison towns against SPLA attacks. In return, the Sudanese government provided Okello's troops with food, shelter, and transport and most likely also provided arms and other military supplies. Then, beginning in August 1986, the UNLA launched operations against NRA units in Kitgum and Gulu districts. Although Museveni closed the border with the Sudan, the attacks continued. In May 1987, the Ugandan and Sudanese governments conducted discussions to improve border security. The following month, Museveni visited Khartoum and held talks with Sudanese prime minister Sadiq al-Mahdi. Among other things, the two leaders agreed that "the Sudan should not allow Ugandan rebels to use its territory as a platform or Uganda to allow Sudanese rebels to remain in its territory."[36] In February 1988, Uganda and the Sudan signed another border security accord, under which both nations promised to cooperate in exposing "criminals and dissidents" in one another's territory.[37]

Despite these agreements, the Ugandan-Sudanese border remained unstable. Anti-Museveni rebels continued to operate from bases in southern Sudan against targets in northern Uganda. Also, the Sudanese government accused the Ugandan president of providing military support to the SPLA. Museveni denied this charge, despite his friendship with SPLA leader John Garang.[38]

In June 1988, the border situation became more complex when the SPLA crossed into Arua and Moyo districts. According to Ugandan government spokesmen, SPLA troops assaulted, kidnapped, and murdered civilians. They also burned and looted several villages, apparently in search of food and other supplies.[39] As a result, many Ugandans sought refuge among Ugandan communities in southern Sudan. To stabilize relations, the Ugandan-Sudanese Permanent Joint Ministerial Commission of Cooperation (PJMCC) met in September 1988 and promised to improve border security and to permit the mutual exchange of refugees.

Although rebels continued to violate the Ugandan-Sudanese border, there was some progress with the refugees. On 7 November 1988, the United Nations High Commissioner for Refugees (UNHCR) announced that it had repatriated 11,000 Ugandans. In addition, the UNHCR declared that Ugandans still in the Sudan could return home in small groups whenever they wished. However, about 15,000 Ugandan refugees who had wanted to return to Uganda remained in the Sudan. Also, about 18,000 Sudanese refugees had settled in northern Uganda around

Adjumani in Moyo. Neither country possessed the resources to provide adequate care for these refugees. Consequently, many people on both sides of the border suffered from a lack of food, shelter, water, and medicine.

In 1989, relations between the two countries took a turn for the worse. Apart from the continuing refugee dilemma, cross-border raids increased. Between 3 and 16 March 1989, Ugandan and Sudanese officials met in Kampala and agreed to contain border incidents and keep all refugees 50 miles from the border.[40] However, both countries lacked the ability and resources to enforce these decisions. Refugees traveled across the border at will. Also, incursions by SPLA rebels continued and Ugandan insurgents still operated out of southern Sudan.[41] Another meeting by the PJMCC failed to achieve any lasting solution to these problems.[42] Relations between the two countries took a turn for the worst on 30 June 1989, when a group of military officers, led by Omar al-Bashir, seized power in the Sudan. The new Sudanese government was determined to end Uganda's support of the SPLA. To achieve this goal, Khartoum sanctioned the use of military force against targets in northern Uganda. Consequently, on 15 November 1989, a Sudanese Air Force (SAF) MiG-19 bombed Moyo, supposedly in reprisal for President Museveni's decision to allow the SPLA to establish bases inside Uganda. Museveni rejected the Sudanese accusations. Three people died in the raid. About six weeks later, a Sudanese army force—composed of sixteen trucks, an armored personnel carrier, and artillery pieces—attacked a Ugandan military post at Oraba and killed one NRA soldier. According to Kampala, pro–Idi Amin rebels, operating from Kaya, Sudan, participated in this operation. Although Khartoum denied having ordered the assault and blamed a local commander for mistakenly having launched the strike, the Ugandan government remained unconvinced. On 1 January 1990, the NRA reoccupied Oraba.[43]

Therefore, President Museveni secured President Omar al-Bashir's promise to establish security monitoring units along the border. Less than two weeks after President Museveni announced this agreement, a Sudanese Air Force jet bombed Moyo again, killing five and injuring six. On 2 April 1990, the two nations signed a nonaggression pact that supposedly guaranteed "that no armed action will be taken by either country against the other." The pact also obliged each country not to allow its territory to be used for launching military operations against the other.[44] To enforce this pact, the Sudan agreed to deploy a nine-man military team to Uganda to enhance security along the common border of the two states. Although it improved political relations, the pact failed to restore stability because the nine-man military team was unable to patrol adequately a border hundreds of miles long.[45]

In November 1991, SPLA personnel, whom Kampala claimed were "deserters," pillaged various areas in Moyo District. Then, on 20 September 1991, two SAF planes bombed a school in Tara, Arua District. Additionally, the refugee situation deteriorated after the SPLA split into two warring factions.[46]

According to the Museveni regime, about 80,000 Sudanese fled into Moyo District to escape fighting between the John Garang and Lam Akol groups.

This exodus continued throughout early 1992, as another 80,000 Sudanese sought refuge from southern Sudan's war in Moyo and Adjumani. Additionally, in January 1992, the UNHCR started returning 2,700 Ugandans who had stayed in the Sudan following the 1988 repatriation.[47] Although they cooperated on these refugee matters, Kampala and Khartoum remained divided over allegations that each provided aid to insurgent groups that operated from bases in the Sudan against Uganda or vice versa. As a result, on 21 May 1992 the Sudan again bombed Moyo to discourage Kampala from providing support to the SPLA.

In mid-1993, the Sudanese government launched a rainy-season offensive against Garang's forces, which caused up to 100,000 Sudanese refugees to pour into northern Uganda. The following year, the number of Sudanese refugees in northern Uganda more than doubled, ensuring that relations between the two countries remained tense. Although President Museveni and President Bashir held a meeting in Vienna to resolve their differences, relations between the two countries remained strained. Kampala also became increasingly alarmed about the prospect of a wider conflict with Khartoum. Although a full-scale war between Uganda and the Sudan is unlikely, ongoing instability in both countries and the absence of any control over the border will keep relations between them tense for the foreseeable future.

Since 1986, Ugandan-Kenyan relations have been strained largely because of personal animosity between Presidents Museveni and Moi.[48] This rivalry stems partially from the fact that the former is a revolutionary who believes force can sometimes achieve political goals whereas the latter is a conservative who cannot tolerate opposition. The inability of both nations to control their common border has also caused several incidents that have given each leader justification for criticizing the other. Lastly, Moi is convinced that Museveni has provided assistance and sanctuary to Kenyan dissidents. Museveni believes that Moi allows anti-NRM elements to operate openly in Kenya.

Shortly after the NRM seized power, the Kenyan government claimed it had discovered links between an anti-Moi group known as Mwakenya and arms smugglers who operated in eastern Uganda. Although Kampala repeatedly denied this accusation, Nairobi remained skeptical.[49] There were also disagreements about illegal immigrants, smuggling, and criminal activities by Ugandans who lived in western Kenya. Eventually, Justus Ole Tipus, the Kenyan minister of state in the office of the president, urged all Ugandans who lived in Kenya to return to their homes and help their kinsmen to reconstruct their country. Few Ugandans accepted this "invitation." Apart from these problems, Moi feared the emergence of a radical Kampala-Tripoli alliance, which would result in the creation of a Jamhariya socialist republic in Uganda.[50] Many Kenyan officials believed that this concern was justified, especially after Libyan leader Muammar Qadhafi arrived in Kampala on 6 September 1986 for a three-day visit on his way back from the Non-Aligned Conference in Harare, Zimababwe. Reportedly, Qadhafi's party, which numbered more than 300, balked at paying its hotel bill, which amounted to Shs. 231 million.

Tensions between the two countries intensified in January 1987, when President Museveni claimed that disloyal Ugandans, operating from bases in Kenya, planned to overthrow his government. Nairobi denied this charge and promised that it would not allow dissidents to operate against Kampala from Kenya. This conciliatory response temporarily improved Ugandan-Kenyan relations. However, in early April 1987, the Ugandan High Commission in Nairobi complained about the alleged harassment of Ugandan nationals who lived in Kenya and Ugandan officials who traveled to Kenya on official duty. The Ugandan government then advised all its citizens to leave Kenya, as it could no longer guarantee their safety. A few weeks later, relations between the two countries deteriorated further after a Kenyan journalist claimed that more than 200 Kenyan dissidents had gone to Libya via Uganda for military training. There also were reports that 1,000 Libyan advisers had deployed to Uganda and had undertaken several projects, including training Kenyan dissidents at Ugandan military installations.

Although there was no evidence to support these charges, Uganda and Kenya remained hostile toward one another. Kenya temporarily shut off telephone service to Uganda, and Kampala discontinued electricity to Nairobi. Museveni also asked Tanzania for oil after a blockade on the Kenyan border had caused a shortage. On 14 December 1987, the two countries reached the brink of war after a series of clashes around Busia, which lasted three days and resulted in military and civilian casualties. In the days following these clashes, the NRA prepositioned heavy military equipment along the border near Busia. Kenya responded by putting its forces on twenty-four-hour alert and deployed General Service Unit personnel to the border to reinforce the police. Kenya also ordered the expulsion of the Ugandan high commissioner and the closure of the Libyan People's Bureau in Nairobi, which reflected the belief that Tripoli was using Uganda to undermine the Moi regime. However, the crisis ended as quickly as it had started. On 28 December 1987, Museveni and Moi met at the border post of Malaba and promised to work for a peaceful solution to the conflict.[51]

Although the two countries eventually concluded an agreement to improve border security, relations between Kampala and Nairobi remained tense. On 16 July 1988, several Ugandan soldiers attacked and robbed Kenyan fishermen at Sumba Island, in Kenyan territory on Lake Victoria.[52] Kenyan security forces deployed to the border island and responded by inflicting some casualties on the Ugandans. Subsequent outbreaks of violence and accusations that Kenya had closed its side of the border prevented a stabilization of the border region.

Problems intensified on 2 March 1989, when some 300 armed Ugandan cattle rustlers entered Kenya's West Pokot District, killed a Kenyan army officer, and injured a civilian. Kenyan security forces responded by killing seventy-two cattle rustlers. Five days later, the Kenyan government claimed that an unidentified military aircraft flew from Uganda, crossed the border near Oropoi, and dropped two bombs near Lokichogio's police post, killing five people and injuring seven others. Kampala denied complicity in the attack and correctly pointed out that the plane had originated in the Sudan.[53] Despite mediation efforts by the Ugandan minister of state for foreign affairs, Tarsis Kabwegyere, the two countries remained at loggerheads.

In 1990, the acrimony between Kampala and Nairobi continued, especially after Ugandan police officials claimed that President Moi had promised to help Ugandan dissidents overthrow President Museveni. On 17 August 1990, the two leaders met in Tororo and agreed to restore full diplomatic ties and to strengthen border security. However, by year's end the two countries had clashed again, largely because of a series of Kenyan newspaper reports that maintained that President Museveni wanted to establish a Pax Uganda on central and eastern Africa.[54]

In early 1991, this war of words degenerated into a shouting match between Presidents Museveni and Moi. Problems started when Museveni announced that Kenya had drawn up plans to recruit 500 foreign mercenaries to invade Uganda and to overthrow his regime. Next, a Ugandan newspaper claimed that Moi had trained members of a Kenyan-based Ugandan exile group called the Ninth October Army/Movement as part of this invasion plan. Moi denied these charges and then accused Museveni of plotting to overthrow his government. The Kenyan government continued the diatribe by proclaiming that Uganda had acquired SCUD surface-to-surface missiles from Iraq via Libya to use in the invasion of Kenya. On 11 February 1991, the semiofficial *Kenya Times* also charged Uganda with undertaking a military buildup along that country's border with Kenya. Over the next few months, tensions remained high as security personnel from both countries clashed at several points along the common border. One such incident occurred on 5–6 April 1991, when Kenyan police shot dead two Ugandan soldiers near Busia. In addition, Nairobi supposedly acquired some Ugandan government documents that indicated that Kampala had recruited thirty-three Kenyan youths for guerrilla training in Libya. A degree of normalcy returned after President Moi made a one-day visit to Uganda on 9 November 1991 and signed a joint communiqué with President Museveni that pledged both countries to "good neighborliness and peace."[55]

In early 1992, this euphoria quickly disappeared after Nairobi alleged that there were Libyan-trained Kenyan youths in Uganda planning to invade Kenya. These charges escalated as Kenya became more unstable as a result of violence associated with that country's democracy movement. Beginning in April 1992, supporters of Kenya's main opposition party, the Forum for the Restoration of Democracy, started fleeing to Uganda to seek asylum. According to many of these refugees, government-inspired violence in western Kenya had forced them to leave their homes.[56] By midyear, the gulf between Uganda and Kenya was as wide as ever.

However, in late 1993 tensions between the two countries eased considerably after President Moi made a three-day visit to Uganda. On 17 November 1993, the two leaders signed an agreement that pledged to expand cooperation in the fields of security, energy, agriculture, education, trade, and transport. By mid-1994, Uganda no longer perceived Kenya as a threat but as a troubled country whose growing internal instability posed a problem to the entire region. Presidents Moi and Museveni also continued to downplay their past personal differences. Nevertheless, as neighbors with a history of acrimony, the two countries are unlikely to effect a permanent solution to their many political and economic problems. However, a resurgence of tension between Uganda and Kenya, which will recur from time to time, will, at the very worst, result in nothing more than low-level border skirmishes, as both nations lack the resources to sustain a large-scale conventional conflict.

The presence of large numbers of Rwandan refugees in Uganda has been a perpetual source of tension between Kampala and Kigali. That many of these refugees supported Idi Amin while he was in power provoked official displeasure and retribution during Obote's second regime. In 1982, Obote, hoping to resolve this refugee problem and prevent challenges to his administration, expelled 60,000 ethnic Rwandans for alleged antigovernment activities. Many of these people were Ugandan citizens whose families had lived in Uganda since the late 1800s.

Even before Museveni, who was of Ankole descent but had relatives in Rwanda, came to power, he recruited some Rwandans into the NRA. As a result, British journalists reported that the Rwandans formed a major element of the original NRA. After he became president in 1986, Museveni recruited approximately 1,000 new Rwandans into the military. This action caused government critics to accuse him of turning the army over to foreigners. Rumors that ethnic Tutsi Rwandans serving in the Ugandan military had formed a group called the Rwanda Patriotic Front alarmed the Hutu-dominated government in Kigali, which believed that the organization posed a threat to Hutu control of Rwanda. Other observers reported that some officials in Kigali believed that Museveni had promised to oust the Rwanda government in exchange for military support while he was leading a guerrilla army in western Uganda.

In 1989, Kampala and Kigali escalated efforts to resolve their differences. In February 1989, for example, the two governments discussed the refugee problem in Uganda. As a result, Kampala agreed to naturalize an unspecified number of Rwandan refugees and Kigali pledged to consider repatriating some Rwandans living in Uganda on a case-by-case basis. In early May 1989, Museveni met with

Rwandan president Juvénal Habyarimana and signed a joint communiqué that affirmed their commitment to resolve the refugee problem with UNHCR assistance. Both governments believed that these discussions marked the beginning of improved Ugandan-Rwandan relations.

Despite this optimism, relations between the two countries soon deteriorated. On 1 October 1990, the RPF invaded Rwanda from bases in Uganda. The initial force, which numbered a few thousand, eventually grew to approximately 7,000. This figure included about 4,000 soldiers who had deserted from the NRA. The other 3,000 were civilians. With few exceptions, most RPF personnel were Rwandan refugees who lived in Uganda. The RPF's Eight-Point Program called for national unity, democracy, an end to corruption and malfeasance, a solution to the refugee problem, a progressive foreign policy, democratization of the armed forces, a greater commitment to social services, and the creation of a self-sustaining economy.[57]

As the war spread throughout northern Rwanda, acrimony between Kampala and Kigali intensified. President Habyarimana repeatedly accused President Museveni of providing bases and covert military assistance to the RPF and of preparing to invade Rwanda, charges that the latter consistently denied. Kampala complained that Rwandan government troops regularly conducted hot pursuit operations into Uganda.[58]

On 11 November 1990, Museveni sought to distance Uganda from the RPF by announcing the retirement of all noncitizen officers and men (namely, Rwandans) from the NRA.[59] This tactic failed to end speculation about a Uganda-RPF alliance, largely because by early 1993 Museveni had failed to act on his promise to expel all Rwandans from the NRA. Moreover, an increasing number of Western military observers doubted that the RPF could sustain military operations without external support.

Although Rwanda and the RPF signed the N'sele cease-fire agreement on 13 July 1992, Kigali continued to suspect the Museveni regime of providing support to the rebels. To minimize tension, Uganda and Rwanda concluded a bilateral security agreement on 8 August 1992 whereby either government could station a military monitoring team on each other's territory to prevent cross-border violations. Approximately one month later, Kampala and Kigali signed an accord to prevent cross-border arms trafficking. However, many Rwandans and Western observers remained convinced that the RPF continued to receive military aid from Uganda, despite the fact that Kampala allowed international military observers to monitor its borders.

In 1993, there was an increase in international pressure for a negotiated settlement to the war, which led to a complex series of negotiations between the RPF and the Rwandan government, sponsored by the Organization of African Unity and Rwanda's neighbors. Tanzania, supported by several African and Western governments, spearheaded this diplomatic effort. Finally, on 4 August 1993, the two warring parties signed the so-called Arusha Accords. Apart from allowing the RPF to join a new national Rwandan army, this agreement allowed the rebels to participate in a new interim government. Also, there were provisions for Rwandan refugees who fled to Tanzania, Uganda, Zaire, and Burundi to return home. To prevent a resurgence of fighting, the United Nations Observer Mission on the Uganda-Rwanda border deployed 105 unarmed peacekeepers to Kabale, which is in southern Uganda near the Rwandan border.

Unbeknownst to the outside world, Hutu extremists secretly planned to wreck the Arusha Accords. After the Rwandan and Burundian presidents died in a mysterious plane crash on 6 April 1994, Hutu fanatics immediately launched a genocidal campaign against the Tutsis and within a matter of months slaughtered 500,000 or more of them. Hundreds of thousands of Tutsis survivors fled to Uganda, Tanzania, and Burundi. As the RPF, reportedly with the support of the Ugandan government, slowly extended its control over Rwanda, approximately 1.5 million Hutus, fearing retribution, fled to eastern Zaire. The resulting humanitarian crisis prompted an international intervention under the umbrella of the United Nations Assistance Mission in Rwanda (UNAMIR). To support this operation, the United States established and managed an "airhead" in Entebbe, which became the focal point of the relief effort to provide food, medicine, and other aid to Rwandan refugees in eastern Zaire.

With an RPF-dominated government in Kigali, Rwandan-Ugandan relations will remain close for the foreseeable future, if for no other reason than the close personal relationship between President Museveni and RPF leader Paul Kigame, who is now Rwandan minister of defence. However, it is unlikely that all of the approximately 200,000 Rwandans in Uganda will return to their homeland, as many of them have established economically prosperous lives in Uganda. As a result, the presence of large numbers of Rwandans in Uganda, nearly all of whom are perceived as foreigners, could complicate future relations between Kampala and Kigali.

International Organizations

Historically, Uganda has been an active member of numerous international organizations. On 25 October 1962, the country joined the UN and within a year had become a member of the International Monetary Fund, the International Bank for Reconstruction and Development, and the International Development Association. Additionally, Uganda was a founding member of the OAU.

Ugandan diplomats have also been active in many other important international organizations, including the Commonwealth, which evolved from the British Empire, and the Non-Aligned Movement (NAM), which seeks to create "a more just, equitable, and peaceful world order."[60] Generally, Uganda has perceived these international organizations as vehicles to acquire foreign economic and technical assistance and to end European minority rule in Africa.

Since 1986, Uganda has played a more active leadership role in these and other international organizations. On 9 July 1990, for example, Yoweri Museveni was elected OAU chairman. During his tenure, he took a number of unprecedented steps that could eventually change the nature of the OAU. On 2 August 1990, Museveni indicated that international military intervention in war-torn Liberia probably was necessary to end that country's civil war. This acknowledgment helped to legitimize the subsequent deployment of peacekeeping troops of the Economic Community of West African States to Liberia. Museveni also supported the establishment of an African Economic Market, which he said would be similar to the European Community. A treaty approved on his last day as OAU chairman laid the groundwork for the creation of this organization, whose goals would be to remove all barriers to inter-African trade and travel and to promote development through economic integration.[61]

In the UN, Uganda worked on several important issues. Apart from pressing for political change in South Africa, the Ugandan government repeatedly urged for the reform of the UN so that it could accommodate the challenges of the 1990s and the next century. In particular, Uganda called for a review of the policy that allowed permanent membership in the Security Council. The Ugandan government also supported the 1990 UN military intervention in Kuwait and in 1992 offered to send troops to Somalia in support of United Nations Operation in Somalia II.

As a result of its activities in the OAU, UN, and other international organizations, Uganda has gained the reputation of being one of the more influential African nations in these organizations. This has helped to further enhance the Museveni regime's prestige among Uganda's neighbors and, more important, among Western donor nations and financial institutions.

Uganda, Refugees, and the East African Arena

Since independence, Uganda has supported all major international initiatives that govern the status of refugees, including the Protocol Relating to the Status of Refugees (1967) and the OAU Convention Governing the Specific Aspects of Refugee Problems in Africa (1969). Although Uganda's constitution contains no reference to refugees, the Control of Alien Refugees Act (1964) makes provision for the care and control of refugee populations in Uganda.

In a region marked by warfare, instability, and political turmoil, Uganda has generated and received hundreds of thousands of refugees. Over the past several decades, Uganda has received large numbers of refugees, primarily from Rwanda, the Sudan, and, to a lesser extent, Zaire, Kenya, and Somalia. At the same time, countless Ugandans have escaped the turmoil caused by a succession of their own leaders by seeking refuge, mainly in the Sudan and Kenya but also in Zaire.

Ugandan president Yoweri Museveni confers with Roger Winter, director of the U.S. Committee for Refugees, about the status of Uganda's many refugees. (Photo courtesy of U.S. Committee for Refugees.)

In early 1994, there were more than 250,000 registered and tens of thousands of unregistered refugees in Uganda. The movement of so many people in and out of Uganda not only has contributed to the country's internal insecurity but also has placed a severe drain on limited Ugandan resources. Nevertheless, historically Uganda has been hospitable to refugees and has normally given them equal access with Ugandans to the country's social services. Kampala has also received help from various international humanitarian agencies such as the UNHCR to help care for refugees or from the U.S. Committee for Refugees to focus world attention on the plight of refugees in Uganda.

7

THE UGANDAN
EXPERIENCE AFTER
THREE DECADES

I T HAS BEEN JUST OVER THREE DECADES since Uganda achieved its independence. During much of that time, the country was in anarchy, suffering from instability, violence, and sheer savagery. There are at least two interpretations to explain this tragedy.

One argues that the destruction associated largely with the regimes of Idi Amin (1971–1979) and Milton Obote (1981–1985) resulted from the country's colonial past.[1] According to this supposition, British imperialists exploited existing ethnic divisions among Ugandans, then implemented a divide-and-rule policy, and finally co-opted the Baganda to act as agents on their behalf in areas outside Buganda. The country remained polarized after independence, thereby setting the stage for the ravages of Amin and Obote, both of whom stayed in power by employing tactics perfected by the British.

The other interpretation maintains that the Ugandans have no one to blame but themselves for the country's political, economic, and social ills. The ethnic divisions that the British supposedly created existed long before the first European ever set foot in Uganda. Moreover, the country's traditional rulers have used violence and intrigue for centuries to attain or preserve political power. Put more simply, this view posits the notion that Ugandans are incapable of managing a modern nation-state or operating a government that functions on democratic principles.[2]

These two interpretations, which represent opposite ends of the political spectrum, provide little, if any, insight into contemporary Uganda. Trying to assign blame for Uganda's present ills on events that occurred decades or even centuries ago reduces history to a propaganda weapon and allows officials and governments to avoid responsibility for their actions. This is not to suggest, however, that historical problems have not plagued Uganda. On the contrary, violence, corruption, ethnic conflict, human rights abuses, weak and ineffective government, social and economic difficulties, and narrowly based political power have characterized precolonial, colonial, and postcolonial Uganda. Nevertheless, the Museveni regime can be judged only according to its own merits and by the success or failure of its efforts to resolve these problems.

Present Policies and Past Problems

President Museveni has sought to break the cycle of violence that has plagued Uganda for much of the postindependence period by employing a "carrot and stick" strategy toward antigovernment rebels. Thus, the authorities use force against its opponents while at the same time extolling them to give up their arms and to rejoin normal society. Although these tactics have reduced the level of lawlessness, they have failed to restore peace and stability to all parts of Uganda. Organized resistance and banditry—which admittedly are low and pose no threat to the government's existence—continue in parts of northern, northeastern, and western Uganda.[3]

Economically, the NRM has made some headway toward rehabilitating the country's infrastructure and material well-being. Despite this success, Kampala remains incapable of implementing the policies necessary to facilitate the kind of development required to eradicate the grinding poverty that has characterized Uganda for much of the postindependence period. The government's chief shortcomings are its inability to abolish corruption, manage the budgetary process, and administer development projects successfully. Uganda's economic self-reliance has been undercut by the country's dependence on a steady infusion of Western largess and technical assistance.

To restore good government, President Museveni devised a grass-roots political strategy that sought to bring about participatory democracy in Uganda. At the heart of this scheme is the NRC, which functions as a legislature, and the RCs, which allow for local control over local affairs, subject to superior laws. Although they have given many Ugandans access to the political system, these institutions are weak and suffer from corruption, favoritism, and incompetence. In addition, hundreds of RC members have been killed by antigovernment rebels who wish to discredit the NRM.

Opponents have challenged President Museveni's political strategy and questioned his commitment to establishing a genuine democratic form of government in Uganda. In particular, they point to his refusal to lift the ban on political parties and his support of "no-party" politics and a "no-party" constitution.[4] Nevertheless, many critics seem willing to reserve final judgment on President Museveni's political aspirations until 1995, when direct presidential elections are scheduled.

Although it has made some progress toward establishing political stability and economic prosperity in parts of Uganda, the NRM has yet to extend the benefits of its policies to many areas of the country. As a result, far too many people have no stake in society and must rely on crime, corruption, or banditry to survive. This situation is unlikely to improve anytime soon, as the Ugandan government lacks the resources to provide an alternative lifestyle for these people.

There are several factors to keep in mind when judging the NRM's performance. When Yoweri Museveni took the oath of office in January 1986, almost all aspects of Ugandan society had been destroyed or severely damaged. The country was also still in the throes of a large-scale war, as numerous guerrilla groups battled government forces and one another. Eventually, the NRM established a modicum of stability, albeit often at the expense of violating human rights. In the areas of social and economic development, the NRM compiled a similar record. Advances have been made in some areas, but the government's accomplishments have been uneven and an overwhelming number of problems continue to impede reconstruction efforts and to tax the NRM's limited resources.

The Path Ahead

Uganda's future prospects are bleak. No matter what policies the NRM or any successor government adopts, the country will remain desperately poor and unstable for the foreseeable future. Moreover, the AIDS epidemic threatens to devastate all levels of Ugandan society, including the Western-educated, skilled class of people who had been expected to help rehabilitate the country. As a result, Kampala's progress toward establishing a peaceful and prosperous society will be limited and largely contingent on the West's willingness to provide ever-growing amounts of aid and technical assistance.

Maintaining access to such aid will depend on the NRM's ability to pursue political and economic policies that are acceptable to the Western donors. Thus, Uganda will be in the difficult position of trying to suppress rebels, bandits, and political opponents while placating the West on issues such as democratization, respect for human rights, a nonconfrontational foreign policy, and market-oriented economic reforms. Additionally, Kampala will have to be reasonably supportive of Western initiatives in the regional and international arenas. Making progress toward achieving these goals, which would benefit all Ugandans, will depend on whether the current and future leadership can avoid a relapse into the self-destructive behavior and fratricidal infighting that has troubled the country since independence.

Preventing such a tragedy will not be easy, especially in view of the fact that Uganda cannot control its borders or maintain internal order in all parts of the country. Armed elements, whether they be foreign soldiers, guerrillas, bandits, cattle raiders, or refugees, move freely in and out of Uganda from neighboring countries and operate with varying degrees of impunity in some areas of Uganda. The resulting instability perpetuates a culture of violence, which is exacerbated by domestic factors such as ethnic tensions, widespread criminal activity, and official corruption. Given such an atmosphere and the government's limited resources, achieving peace and prosperity are not realistic goals. Instead, Kampala must seek to minimize rather than try to eliminate the causes of this disorder.

Relying on military force to achieve this goal is counterproductive because the country's chronic instability results not from traditional security threats but rather from an array of political, economic, and social problems. To resolve these difficulties, the Museveni regime or any successor government must concentrate on creating a unified Ugandan nation out of a conglomeration of different, often warring, ethnic groups.

The Ugandan government must also supplement its nation-building efforts with more energetic and focused policies designed to facilitate more equitable distribution of the country's political, social, and economic resources. With a few exceptions, most rehabilitation and development work has been concentrated in the south. Whether by design or coincidence, this emphasis on southern well-being helps to perpetuate ethnic and regional animosities.

The NRM could lessen these tensions and boost its popularity by designing schemes to better demonstrate its commitment to ensuring that all Ugandans benefit equally from government development policies. In the economic arena, for example, the $179 million Northern Uganda Reconstruction Program received considerable attention as an indication of Kampala's determination to rebuild the war-torn north. In mid-February 1993, after a long delay because of guerrilla activity in Gulu District and elsewhere, the government announced the official opening of this program, which was financed by the World Bank, Denmark, and the Netherlands. However, by mid-1994 this program had made little progress toward improving the lives of the northerners, many of whom still lived grim lives.

Apart from the fact that similar comprehensive economic development strategies need to be devised for eastern and western Uganda, the Ugandan government and Western donors must ensure that such programs are implemented in a timely fashion. This is not to suggest that the NRM has totally neglected these regions.

On the contrary, the Ugandan government has undertaken numerous development projects in eastern and western Uganda. On 29 December 1992, for example, Kampala signed a $3.9 million agreement with the UNDP to alleviate poverty and to facilitate rural development in West Nile. However, such programs, though important, will not generate the local enthusiasm that accompanied the announcement of the Northern Uganda Reconstruction Program. Even if it failed to acquire adequate international funding for comprehensive regional development programs, the NRM undoubtedly would enhance its credibility among eastern and western Ugandans who believe that President Museveni has not delivered on promises to improve their lives.[5]

Politically, the RC system has allowed thousands of Ugandans to participate in the governing process. However, President Museveni's refusal to lift the ban on political parties undercuts his promises to facilitate the establishment of a democratic government in Uganda. Moreover, his support of such dubious concepts as "no-party" politics and a "no-party" constitution provides opponents with ample justification to criticize him and the NRM for duplicity and chicanery.

Unless he comes to terms with growing demands for genuine democracy, President Museveni faces a future in which he will be increasingly identified as a repressive, autocratic leader who cares more about preserving his own position than improving Uganda's political system. However, if he sanctions multipartyism, President Museveni could run the risk of fostering politically motivated violence and instability. Whatever path he (or any of his successors) pursues, it is likely that Ugandan democracy will be similar to democratic systems all over the world insofar as it will be fragile, flawed, and fractious.

The last, and perhaps most significant, factor that must be taken into account when assessing Uganda's future prospects is Yoweri Museveni. His tenure as president has been marked by numerous political, economic, and military successes and failures. However, his most enduring legacy does not concern bureaucratic battles won and lost; rather, his place in history will be determined largely by the type of government he leaves behind after relinquishing the presidency.

Unlike Kenya and Tanzania, both of which have experienced peaceful transfers of presidential power, Uganda has endured much hardship and violence during changes of government. Despite this turbulent past, Museveni has failed to groom a successor or create the machinery necessary to facilitate a smooth transition to a new regime. As a result, Uganda could very well experience a vicious power struggle after he leaves the political stage. Such a development would almost certainly exacerbate ethnic tensions and would encourage the use of force by groups seeking political power. These are precisely the forces that have caused so much havoc and destruction in Uganda.

For the country to avoid yet another apocalypse, President Museveni must look beyond himself to the Uganda of the future. His energies should be devoted to ensuring that a nonviolent transition process is established, that the NRM's dream of a peaceful and prosperous Uganda endures, and that all Ugandans disabuse the notion that violence should be used to gain or preserve political power.

Admittedly, such aspirations are grandiose and seemingly out of place in Africa, especially for a country such as Uganda, whose history has been so troubled and violent. Moreover, any progress toward achieving these goals will inevitably be fraught with setbacks and partial achievements. Nevertheless, there is no alternative to pursuing these dreams if President Museveni is to honor his promise to ensure that all Ugandans "regain their pride and confidence" and that Uganda assumes "its right place in the civilized world."[6]

Notes

Chapter 1

1. For more detailed descriptions of Uganda's mineral resources, see H. B. Thomas and Robert Scott, *Uganda* (London: Oxford University Press, 1935), pp. 212–221; and Allison Butler Herrick et al., *Uganda: A Country Study* (Washington, D.C.: USGPO, 1981), pp. 22–24, 279–281.

2. Herrick et al., *Uganda: A Country Study*, pp. 23–24.

3. Ministry of Planning and Economic Development, *Background to the Budget, 1991–1992* (Kampala: Ministry of Planning and Economic Development, 1991), p. 95.

4. This section is based on Herrick et al., *Uganda: A Country Study*, p. 15; and Thomas and Scott, *Uganda*, pp. 115–116.

5. *New Vision* (23 November 1989), p. 4.

6. This section is based on Herrick et al., *Uganda: A Country Study*, pp. 16–22.

7. Today, such areas are known as forest reserves.

8. Thomas and Scott, *Uganda*, p. 157.

9. A comprehensive review of Uganda's trees can be found in W. J. Eggeling (revised and enlarged by Ivan R. Dale), *The Indigenous Trees of the Uganda Protectorate*, 2d ed. (Entebbe: Government Printer, 1952). Another useful work is W. J. Eggeling, *Notes on the Forests of Uganda and Their Products* (Entebbe: Government Printer, 1947).

10. For an assessment of Uganda's forest problems, see A. C. Martin, *Deforestation in Uganda* (Nairobi: Oxford University Press, 1984); and Peter C. Howard, *Nature Conservation in Uganda's Tropical Forest Reserves* (Gland: International Union for Conservation of Nature and Natural Resources, 1991).

11. *New Vision* (31 July 1989), p. 4.

12. *African Economic Digest* (9 September 1991), p. 11.

13. The Danish Agency for International Development (DANIDA) funded this project, which aims to establish 3,950 tree nurseries in all the country's thirty-eight districts.

14. *Africa Economic Digest* (4 May 1992), p. 10. Also see *New Vision* (9 April 1992), pp. 1, 16; and (13 April 1992), p. 16.

15. For an early discussion of the man-animal conflict, see Uganda, *Game Preservation and Economic Development* (Entebbe: Government Printer, 1928).

16. *New Vision* (10 May 1991), p. 4.

17. Ibid., (20 May 1991), p. 10; also see (18 July 1991), p. 4.

18. *Financial Times* (21 May 1990), p. 1.

19. For additional information, see *New Vision* (10 December 1991), pp. 1, 16.

20. Ibid. (5 December 1991), p. 16.

21. *African Business* (May 1993), p. 41.

22. Ibid. (August 1990), p. 48.

23. Ibid. (October 1991), p. 47.

24. *New Vision* (13 April 1991), pp. 8–9.

Chapter 2

1. Standard accounts of the Lwoo include J. P. Crazzolara, "The Lwoo People," *Uganda Journal*, Vol. 5, No. 1 (1937), pp. 1–21; and J. P. Crazzolara, *The Lwoo*, 3 vols. (Verona: Missioni Africane, 1950–1954). For the classic accounts of Bunyoro, see John Beattie, *The Nyoro State* (Oxford: Clarendon Press, 1971); John Roscoe, *The Bakitara or Banyoro* (Cambridge: Cambridge University Press, 1923); A. R. Dunbar, *A History of Bunyoro-Kitara* (Nairobi: Oxford University Press, 1970); Godfrey N. Uzoigwe, *Revolution and Revolt in Bunyoro-Kitara* (London: Longmans, 1970); and Benjamin C. Ray, *Myth, Ritual, and Kingship in Buganda* (New York and Oxford: Oxford University Press, 1991).

2. For a basic survey, see M.S.M. Semakula Kiwanuka, *A History of Buganda: From the Foundation of the Kingdom to 1900* (New York: Africana Publishing Corporation, 1972).

3. Roland Oliver and Gervase Mathew (eds.), *History of East Africa*, vol. 1 (Oxford: Clarendon Press, 1963), p. 189.

4. H. M. Stanley, *Through the Dark Continent*, vol. 1 (New York: Harper and Brothers, 1878), pp. 305–306.

5. For a traditional history of Ankole, see S. Karugire, *A History of Nkore in Western Uganda to 1896* (Oxford: Clarendon Press, 1971). During the precolonial period, Ankole was called Nkore.

6. Robert M. Maxon, *East Africa: An Introductory History* (Morgantown: West Virginia University Press, 1986), p. 88.

7. Ibid., pp. 88–89.

8. The standard work is Kenneth Ingham, *The Kingdom of Toro in Uganda* (London: Methuen and Company, 1975).

9. David Livingstone's African journeys and the exploits of the German missionaries Ludwig Krapf and Johannes Rebmann aroused the interest of the Royal Geographical Society, which supported the Burton and Speke and the Speke and Grant expeditions.

10. Useful accounts of these two important explorers include J.N.L. Baker, "Sir Richard Burton and the Nile Sources," *Uganda Journal*, Vol. 12, No. 1 (March 1948), pp. 61–71; and J.N.L. Baker, "John Hanning Speke," *Geographical Journal*, Vol. 128, Part 4 (1962), pp. 385–388.

11. For Burton's account of this journey, see Richard F. Burton, *The Lake Regions of Central Africa* (London: Longman, Green, Longman and Roberts, 1860). An assessment of the Burton-Speke dispute is contained in Charles T. Beke, *Who Discovered the Sources of the Nile?* (London: Williams and Norgate, 1863).

12. For accounts of their travels, see John Hanning Speke, *Journal of the Discovery of the Source of the Nile* (New York: Harper and Brothers, 1868); John Hanning Speke, *What Led to the Discovery of the Source of the Nile* (Edinburgh and London: William Blackwood and Sons, 1863); and James Augustus Grant, *A Walk Across Africa* (London: Blackwood, 1864). For a critical analysis of the Speke-Grant explorations, see "Sources of the Nile," *Temple Bar*, Vol. 9 (1863), pp. 108–119. More recent accounts of the two explorers are contained in Alan Moorehead, *The White Nile* (New York: Harper and Row, 1971), pp. 53–74; and Christopher Hibbert, *Africa Explored* (New York and London: W. W. Norton and

Company, 1982), pp. 213–217. An article in the *Times* (19 September 1864) supported the view that Speke had discovered the source of the Nile.

13. For Murchison's views on the Burton-Speke explorations, see Roderick I. Murchison, "Discoveries of Burton and Speke," *Proceedings of the Royal Geographical Society,* Vol. 3 (1858–1859), pp. 301–308. *Journal of the Royal Geographical Society,* Vol. 25 (1865), pp. clxxvi–clxxviii. For Livingstone's assessment of this trip, see Horace Waller (ed.), *The Last Journals of David Livingstone* (New York: Harper and Brothers, 1875); Stanley, *Through the Dark Continent,* 2 vols.

14. For Baker's account of this journey, see Samuel White Baker, *Ismailia: A Narrative of the Expedition to Central Africa for Suppression of the Slave Trade,* 2 vols. (London: Macmillan and Company, 1868).

15. Maxon, *East Africa,* p. 129.

16. For an account of Emin Pasha's life, see G. Schweitzer, *The Life and Work of Emin Pasha,* 2 vols. (London: Constable, 1898). Stanley's account of the expedition is contained in Henry Morton Stanley, *In Darkest Africa,* 2 vols. (London: Sampson Low and Company, 1890). For a critical assessment of Stanley's leadership, see Walter George Barttelot, *The Life of Edmund Musgrave Barttelot* (London: Richard Bentley and Son, 1890).

17. The CMS party, which started out from Bagamoyo, included eight members. However, only two people, Shergold Smith and the Reverend C. T. Wilson, survived the journey and reached Buganda. For a discussion of the CMS goals in Buganda, see A. T. Matson, "The Instructions Issued in 1876 and 1878 to the Pioneer CMS Parties to Karagwe and Uganda," *Journal of Religion in Africa,* Vol. 12, No. 3 (1981), pp. 192–237; and Vol. 13, No. 1 (1982), pp. 25–46. For a more general treatment of this subject, see J. V. Taylor, *The Growth of the Church in Buganda* (London: S.C.M. Press, 1958). Rubaga was part of what became the modern city of Kampala.

18. The 1886 Anglo-German Agreement only provided for East Africa's division up to Lake Victoria's eastern shore.

19. For an assessment of the IBEAC's performance, see John S. Galbraith, *Mackinnon and East Africa, 1878–1895: A Study in the "New Imperialism"* (Cambridge: Cambridge University Press, 1972). For Jackson's account, see Sir Frederick J. Jackson, *Early Days in East Africa* (London: Edward Arnold, 1930).

20. For Peters's account, see Carl Peters, *New Light on Dark Africa* (London: Ward Lock, 1891). When an indignant Peters learned of the Anglo-German agreement, he accused Berlin of sacrificing two African kingdoms (Witu and Buganda) for a "bathtub in the North Sea." Peters's reaction is quoted in Zoë A. Marsh and G. W. Kingsnorth, *An Introduction to the History of East Africa* (Cambridge: Cambridge University Press, 1965), p. 109.

21. For a full text of the agreement, see E. Hertslet, *The Map of Africa by Treaty,* 3d ed. vol. 3 (London: HMSO, 1909), pp. 948–950.

22. Essential to understanding this period is Frederick D. Lugard, *The Rise of Our East African Empire,* 2 vols. (London: Blackwood and Sons, 1893). Interested readers also should consult Frederick D. Lugard, *The Story of the Uganda Protectorate* (London: Horace Marshall and Son, [1900]); and *Reports by Captain F. D. Lugard on His Expedition to Uganda* (London: Doherty, 1890, 1891, and 1892).

23. Kasagama fled into exile after Kabarega of Bunyoro attacked Toro. To regain his throne, Kasagama signed a treaty with Lugard on 14 August 1891 whereby he relinquished

control of Toro and all its dependencies to the IBEAC. Additional information is contained in Ingham, *The Kingdom of Toro in Uganda*, pp. 62–74.

24. Tension between Catholics, and Protestants (Fransa and Ingleza, as they were known at the time) predated Lugard's arrival in Uganda. In the 1880s, competition between Catholic missionaries who favored Germany and Protestant missionaries who identified with the British often prompted fighting between the two groups. In December 1891, with religious tension between Catholics and Protestants still high, Lugard ignored orders from the IBEAC to withdraw from Buganda. Instead, he relied on CMS financial aid to continue his activities in Buganda. When fighting broke out between Catholics and Protestants, Lugard supported the latter.

25. For a discussion of the trip, see Sir Gerald Portal, *The British Mission to Uganda in 1893* (London: Edward Arnold, 1894).

26. The collaboration-resistance theme is common to most literature about Uganda. Some of the better-known works include Edward I. Steinhart, *Conflict and Collaboration: The Kingdoms of Western Uganda, 1890–1907* (Princeton: Princeton University Press, 1977); and J. W. Nyakatura, *Anatomy of an African Kingdom* (New York: NOK Publishers, 1973), pp. 153–171.

27. On 3 December 1897, a 27th Bombay Light Infantry battalion, numbering 14 British officers and 743 Indian personnel, embarked at Bombay on board the transport *Nowshera*. Eventually, the 27th lost 2 Indian officers, 1 lance naik, and 18 privates in battles against the mutineers.

28. The mutiny resulted in 280 deaths and 555 wounded on the government side. An estimated 400–500 Sudanese lost their lives during the mutiny. The authorities also hanged 12 Sudanese after the mutiny had ended. Although it is impossible to give a comprehensive selection of references about the mutiny, interested readers should consult J. V. Wild, *The Uganda Mutiny* (London: Macmillan, 1954). For a more sympathetic account of the Arab cause, see Ibrahim Elzein Soghayroun, *The Sudanese Muslim Factor in Uganda* (Khartoum: Khartoum University Press, 1981).

29. For a survey of the Uganda Agreement of 1900, see John Vernon Wild, *The Story of the Uganda Agreement* (Nairobi: East African Literature Bureau, 1950).

30. Dissidents based their rebellion on the slogan "*Nyangire Abaganda*" ("I have refused the Ganda"). The best account of this rebellion is G. N. Uzoigwe, "The Kyanyangire, 1907: Passive Revolt Against British Overrule," in Bethwell A. Ogot (ed.), *War and Society in Africa* (London: Frank Cass, 1972), pp. 179–214. Also, see Maxon, *East Africa*, pp. 150–151.

31. Kigezi District bordered on the Belgian Congo (now Zaire) and German East Africa (now Tanzania).

32. The classic account is A. D. Roberts, "The Sub-Imperialism of the Baganda," *Journal of African History*, Vol. 3, No. 3 (1962), pp. 435–450. Also see Maxon, *East Africa*, p. 152.

33. For a survey of Uganda's military establishment, see Thomas P. Ofcansky, "National Security," in Rita Byrnes (ed.), *Uganda: A Country Study* (Washington, D.C.: USGPO, 1992), pp. 195–236.

34. Allison Butler Herrick et al., *Uganda: A Country Study* (Washington, D.C.: USGPO, 1981), p. 392.

35. The most detailed account of the 1900 Uganda Agreement is contained in D. Anthony Low and R. Cranford Pratt, *Buganda and British Overrule, 1900–1955* (London: Oxford University Press, 1960), pp. 3–159. Interested readers also should consult A. B.

Mukwaya, *Land Tenure in Buganda* (Kampala: East African Institute of Social Research, 1953); Henry W. West, *The Transformation of Land Tenure in Buganda Since 1896* (Leiden: Afrika Studiecentrum, 1971); and Henry W. West, *Land Policy in Buganda* (Cambridge: Cambridge University Press, 1972).

36. Low and Pratt, *Buganda and British Overrule*, pp. 234–236.

37. Henry E. Colvile, who served as commissioner of Uganda (1893–1895) commanded the expedition against Bunyoro. For his account of the fighting, see Henry E. Colvile, *The Land of the Nile Springs* (London: Edward Arnold, 1895).

38. Part of the reason for Bunyoro's emotional attitude toward the "lost counties" concerns the burial places of the *omukamas*, which are considered to be national shrines. Except for the tombs of two *omukamas* who died after 1900 all other *omukama* graves were in the lost counties. For a brief discussion of this problem, see A. D. Roberts, "The Lost Counties of Bunyoro," *Uganda Journal*, Vol. 26, No. 2 (1962), pp. 194–199.

39. The appropriate documents include Great Britain, *Papers Relating to the Question of the Closer Union of Kenya, Uganda and the Tanganyika Territory* (London: HMSO, 1931); and Great Britain, *Joint Select Committee on Closer Union in East Africa* (London: HMSO, 1931).

40. For a discussion of Uganda's role in World War II, see Major E. F. Whitehead, "A Short History of Uganda Military Units Formed During World War II," *Uganda Journal*, Vol. 14, No. 1 (March 1950), pp. 1–14; and J. C. Worker, "With the 4th (Uganda) K.A.R. in Abyssinia and Burma," *Uganda Journal*, Vol. 12, No. 1 (March 1948), pp. 52–56.

41. Jan Jelmert Jorgensen, *Uganda: A Modern History* (New York: St. Martin's Press, 1981), pp. 181–182.

42. George Delf, *Asians in East Africa* (London: Oxford University Press, 1963), p. 2.

43. Robert G. Gregory, *India and East Africa: A History of Race Relations Within the British Empire 1890–1939* (Oxford: Clarendon Press, 1971), pp. 393–394.

44. Ibid., pp. 486–488.

45. H. S. Morris, *The Indians in Uganda* (Chicago: University of Chicago Press, 1968), p. 109.

46. Allison Butler Herrick et al., *Uganda: A Country Study* (Washington, D.C.: USGPO, 1981), p. 266.

47. African racism toward Indians has a long history in Uganda. Beginning in 1919, the Young Baganda Association launched an anti-Indian campaign throughout the protectorate. Among other things, the Young Baganda Association claimed that Indians were blocking African upward mobility.

48. Two of the better accounts of the Ugandan economy during the colonial period are Cyril Ehrlich, "The Uganda Economy, 1903–1945," in Vincent Harlow and E. M. Chilver (eds.), assisted by Alison Smith, *History of East Africa*, vol. 2 (Oxford: Clarendon Press, 1976), pp. 395–475; and D. A. Lury, "Dayspring Mishandled? The Uganda Economy, 1945–1960," in D. A. Low and Alison Smith (eds.), *History of East Africa*, vol. 3 (Oxford: Clarendon Press, 1976), pp. 212–250.

49. Ehrlich, "The Uganda Economy, 1903–1945," p. 397. Between 1893 and 1915, when they stopped, grants-in-aid totaled £2.5 million.

50. Ibid., pp. 406, 422.

51. Hesketh H. Bell, *Report on the Introduction and Establishment of the Cotton Industry in the Uganda Protectorate* (London: HMSO, 1909).

52. Ehrlich, "The Uganda Economy, 1903–1945," p. 422.

53. World War I left the few European-owned plantations without adequate supervision of the African peasants who performed the day-to-day work. As a result, plantation production declined. It is interesting to note that by the early 1920s, European plantation agriculture had become unprofitable because of the international collapse of coffee and rubber prices.

54. J. D. Tothill (ed.), *Agriculture in Uganda* (London: Oxford University Press, 1940), pp. 101–110.

55. For an analysis of Uganda's post–World War II economic performance, see Lury, "Dayspring Mishandled?" pp. 212–250.

56. Sir John Hall, "Some Aspects of Economic Development in Uganda," *African Affairs*, Vol. 51, No. 203 (April 1952), pp. 124–132.

57. Lury, "Dayspring Mishandled?" pp. 236–238. Also see Walter Elkan and Gail G. Wilson, "The Impact of the Owen Falls Hydro-Electric Project on the Economy of Uganda," *Journal of Development Studies*, Vol. 3, No. 4 (1967), pp. 387–404; and Gail Wilson, *Owen Falls: Electricity in a Developing Country* (Nairobi: East African Publishing House, 1967). It is interesting to note that in mid-1990, the Ugandan government announced plans to build a new hydroelectric power station at Owen Falls. For a description of this project, see *New Vision* (30 July 1990), pp. 1, 16; and (31 July 1990), p. 12.

58. A. B. Adimola, "Uganda: The Newest 'Independent,'" *African Affairs*, Vol. 62, No. 249 (October 1963), p. 327.

59. *The Economic Development of Uganda* (Baltimore: Johns Hopkins Press, 1962), pp. 16, 25, and 37.

60. For the official view of the 1949 riots, see *Commission of Inquiry into Civil Disturbances in Buganda in April and May 1949* (Entebbe: Government Printer, 1950).

61. For the standard account of Ugandan political parties, see Donald A. Low, *Political Parties in Uganda, 1949–62* (London: University of London Institute of Commonwealth Affairs, 1962).

62. Maxon, *East Africa*, pp. 223–224.

63. Ibid., p. 224.

64. Great Britain, *Report of the Uganda Constitutional Conference, 1961* (London: HMSO, 1961).

65. The movement's leaders repeatedly denied that the Kabaka Yekka was a political party. For a discussion of the movement, see Cherry Gertzel, "Report from Kampala," and "How Kabaka Yekka Came to Be," *Africa Report* (October 1964), pp. 3–13; and I. R. Hancock, "Patriotism and Neo-Traditionalism in Buganda: The Kabaka Yekka ("The King Alone") Movement, 1961–62," *Journal of African History*, Vol. 11, No. 3 (1970), pp. 419–434.

Chapter 3

1. Standard biographical studies of Obote include A.G.G. Gingyera-Pinycwa, *Apolo Milton Obote and His Times* (New York: NOK Publishers, 1978); Vijay Gupta, *Obote: Second Liberation* (New Delhi: Vikas Publishing House, 1983); and Kenneth Ingham, *Obote: A Political Biography* (London and New York: Routledge, 1994).

2. Jan Jelmet Jorgensen, *Uganda: A Modern History* (New York: St. Martin's Press, 1981), p. 219.

3. Ibingira is one of Uganda's most experienced politicians. Apart from serving as UPC secretary-general, he was a member of parliament, held various ministerial positions, and

was Uganda's ambassador to the United Nations. From 1966 to 1971, Ibingira, who had fallen afoul of Obote, was in detention. In 1994, Ibingira was in private business in Kampala and was honorary consul of Spain. For his view of Ugandan history, see G.S.K. Ibingira, *The Forging of an African Nation* (New York: Viking Press, 1973).

4. Amin was one of independent Uganda's most important personalities. Some of the more useful biographies include David Martin, *General Amin* (London: Faber and Faber, 1974); Thomas and Margaret Melady, *Idi Amin Dada: Hitler in Africa* (Kansas City: Sheed Andrews and McMeel, 1977); Judith Listowel, *Amin* (London: Irish University Press, 1974); and David Gwyn, *Idi Amin* (Boston: Little, Brown and Company, 1977). Another useful assessment of Amin's career is contained in Samuel Decalo, *Psychoses of Power: African Personal Dictatorships* (Boulder: Westview Press, 1989), pp. 77–127.

5. Martin, *General Amin*, p. 115; Decalo, *Psychoses of Power*, p. 90; and Crawford M. Young, "The Obote Revolution," *Africa Report* (June 1966), pp. 8–14.

6. The most important accounts of this period of Ugandan history include Mutesa II, *Desecration of My Kingdom* (London: Constable, 1967); Milton Obote, *Myths and Realities* (Kampala: African Publishers, 1970); and Akena Adoko, *Uganda Crisis* (Kampala: African Publishers, 1970).

7. For a critical assessment of this program, see Irving Gershenberg, "Slouching Towards Socialism: Obote's Uganda," *African Studies Review*, Vol. 15, No. 1 (April 1972), pp. 79–95.

8. Milton Obote, *The Common Man's Charter, with Appendices* (Entebbe: Government Printer, 1970). Also see Tertit Aasland, *On the Move-to-the-Left, 1969–1971* (Uppsala: Scandinavian Institute of African Studies, 1974). For additional information on the five documents, see James H. Mittelman, *Ideology and Politics in Uganda: From Obote to Amin* (Ithaca and London: Cornell University Press, 1975), pp. 120–130; or Jorgensen, *Uganda*, pp. 234–235.

9. For additional information on the coup, see Michael Twaddle, "The Amin Coup," *Journal of Commonwealth Political Studies*, Vol. 10, No. 2 (July 1972), pp. 99–121; Michael F. Lofchie, "The Uganda Coup-Class Action by the Military," *Journal of Modern African Studies*, Vol. 10, No. 1 (May 1972), pp. 19–35; Ruth First, "Uganda: The Latest Coup d'État in Africa," *World Today*, Vol. 27, No. 3 (March 1971), pp. 131–138; and Aidan Southall, "General Amin and the Coup: Great Man or Historical Inevitability?" *Journal of Modern African Studies*, Vol. 13, No. 1 (March 1975), pp. 85–105. Also see Robert M. Maxon, *East Africa: An Introductory History* (Morgantown: West Virginia University Press, 1986), p. 259; and Amii Omara-Otunnu, *Politics and the Military in Uganda, 1890–1985* (New York: St. Martin's Press, 1987), p. 90.

10. Martin, *General Amin*, p. 14.

11. Major Iain Grahame, "Uganda and Its President," *Army Quarterly and Defence Journal*, Vol. 104, No. 4 (July 1974), pp. 480–481.

12. Henry Kyemba, *A State of Blood: The Inside Story of Idi Amin* (New York: Ace Books, 1977), p. 108.

13. Ibid., p. 39; Jorgensen, *Uganda*, p. 271; and Mahmood Mamdani, *Imperialism and Fascism in Uganda* (Nairobi: Heinemann Educational Books, 1983), p. 37.

14. A copy of the Eighteen Points is contained in Semakula Kiwanuka, *Amin and the Tragedy of Uganda* (Munich and London: Weltforum Verlag, 1979), pp. 40–42.

15. Omara-Otunnu, *Politics and the Military in Uganda*, p. 105; and Decalo, *Psychoses of Power*, pp. 98–99.

16. Amin's excesses are well documented. Some of the more useful accounts include Amnesty International, *Human Rights in Uganda* (London: Amnesty International, 1978); International Commission of Jurists, *Violations of Human Rights and Rule of Law in Uganda* (Geneva: International Commission of Jurists, 1974); and International Commission of Jurists, *Uganda and Human Rights* (Geneva: International Commission of Jurists, 1977).

17. The literature about the expulsion of the Asians is extensive. Some of the more important items include Michael Twaddle (ed.), *Expulsion of a Minority: Essays on Ugandan Asians* (London: Athlone Press, 1975); Thomas and Margaret Melady, "The Expulsion of the Asians from Uganda," *Orbis*, Vol. 19, No. 4 (Winter 1976), pp. 1600–1620; H. Patel, "General Amin and the Indian Exodus from Uganda," *Issue*, Vol. 2, No. 4 (Winter 1972), pp. 12–22; Justin O'Brien, "General Amin and the Ugandan Asians: Doing the Unthinkable," *Round Table*, No. 249 (January 1973), pp. 91–104; and Justin O'Brien, *Brown Britons: The Crisis of the Ugandan Asians* (London: Runnymede Trust, 1972). Also see Uganda, *The Action Programme, 1977/78–1979/80* (Entebbe: Ministry of Planning and Economic Development, 1977), p. 46.

18. Samuel Decalo, *Coups and Army Rule in Africa*, 2d ed. (New Haven and London: Yale University Press, 1990), p. 175; Uganda, *The Action Programme*, p. 31. Also see Commonwealth Team of Experts, *The Rehabilitation of the Economy of Uganda*, vol. 2 (London: Commonwealth Secretariat, 1979), pp. 133–148, 167.

19. For a more detailed discussion of this subject, see Susan Aurelia Gitelson, "Major Shifts in Recent Ugandan Foreign Policy," *African Affairs*, Vol. 76, No. 304 (July 1977), pp. 359–380.

20. An account of the arrest and subsequent developments is contained in Denis Hills, *The White Pumpkin* (New York: Grove Press, 1975), pp. 326–344. Also see *Facts on File* (31 July 1976), p. 375.

21. For a pro-Israeli perspective on this event, see "How Much Did Libya Pay?" *Jewish Observer and Middle East Review* (31 March 1972), pp. 4–5.

22. Ibid. Also see Colin Legum (ed.), *Africa Contemporary Record: Annual Survey and Documents, 1974–1975* (New York: Africana Publishing Company, 1975), p. B-319; and Colin Legum (ed.), *Africa Contemporary Record: Annual Survey and Documents, 1975–1976* (New York: Africana Publishing Company, 1976), p. B-363.

23. Gad W. Toko, *Intervention in Uganda: The Power Struggle and Soviet Involvement* (Pittsburgh: University Center for International Studies, University of Pittsburgh, 1979), pp. 69–76.

24. Ibid., pp. 81–83.

25. For a discussion of Amin's impact on this area, see Marcelino Kombo, "Amin's Pillage in the Kagera," *Africa* (January 1979), pp. 12–17.

26. The best account of this conflict remains Tony Avirgan and Martha Honey, *War in Uganda: The Legacy of Idi Amin* (Westport, Conn.: Lawrence Hill and Company, 1982). For Tanzania's view of the war, see *Tanzania and the War Against Amin's Uganda* (Dar es Salaam: Government Printer, 1979).

27. Rodger Yeager, *Tanzania: An African Experiment*, 2d ed. (Boulder: Westview Press, 1989), p. 136.

28. Quoted in Colin Legum (ed.), *Africa Contemporary Record: Annual Survey and Documents, 1978–1979* (New York: Africana Publishing Company, 1980), p. B-436.

29. Jorgensen, *Uganda*, p. 332.

30. Colin Legum (ed.), *Africa Contemporary Record: Annual Survey and Documents, 1979–1980* (New York: Africana Publishing Company, 1981), pp. B-354–B-355.

31. Ibid, p. B-355; and Francis A. W. Bwengye, *The Agony of Uganda: From Idi Amin to Obote* (London and New York: Regency Press, 1985), p. 74.

32. For a more detailed discussion of the Binaisa regime, see Bwengye, *The Agony of Uganda*, pp. 41–53.

33. Quoted in Phares Mutibwa, *Uganda Since Independence* (Trenton, N.J.: Africa World Press, 1992), p. 133. Also see Don Kabeba, "Special Courts for 'Aminists,'" *New African* (March 1980), p. 65.

34. Legum (ed.), *Africa Contemporary Record, 1979–1980*, p. B-362.

35. Ibid., pp. B-346, B-358, B-362.

36. Mutibwa, *Uganda Since Independence*, p. 134. It is interesting to note that during the early 1990s, this was the same argument made by Yoweri Museveni to justify his "no-party" democracy.

37. Jorgensen, *Uganda*, pp. 335–336.

38. For an official account of the electoral results, see *Report of the Electoral Commission 1980 Presented to His Excellency Dr. A. Milton Obote MP, President of the Republic of Uganda* (Entebbe: Government Printer, 1981). A Commonwealth observer team maintained that the elections had been reasonably free and fair under the circumstances.

39. Colin Legum (ed.), *Africa Contemporary Record, 1981–1982: Annual Survey and Documents* (New York and London: Africana Publishing Company, 1983), p. B-298. Also see Jimmy K. Tindigarukayo, "Uganda, 1979–85: Leadership in Transition," *Journal of Modern African Studies*, Vol. 26, No. 4 (1988), p. 615.

40. James Tumusiime (ed.), *Uganda: 30 Years, 1962–1992* (Kampala: Fountain Publishers, 1992), p. 63.

41. Tindigarukayo, "Uganda 1979–85," p. 616.

42. T. P. Lanser, "Can Obote Survive?" *Africa Report* (January–February 1982), p. 43.

43. *New York Times* (25 February 1982), p. 9A.

44. Ed Hooper and Louise Pirouet, *Uganda* (London: Minority Rights Group, 1989), p. 9.

45. *Christian Science Monitor* (27 January 1986), p. 9.

46. Reported in Catharine Watson, *Exile from Rwanda: Background to Invasion* (Washington, D.C.: U.S. Committee for Refugees, 1991), p. 11. Also see Amnesty International, *Uganda: Six Years After Amin* (London: Amnesty International, 1985), p. 24.

47. Watson, *Exile from Rwanda*, p. 10. As Watson correctly points out, Obote's persecution of the Banyarwanda caused many of them to join the NRA. Indeed, by 1984 the Banyarwanda were the third-largest group in the NRA after the Banyankole and Baganda. By the time it entered Kampala on 26 January 1986, the NRA had about 14,000 personnel, 2,000–3,000 of which were Banyarwanda. In return for their support, Yoweri Museveni supposedly promised the Banyarwanda that he would help them return to Rwanda. On 1 October 1990, the Rwanda Patriotic Front (RPF), which reportedly contained 4,000 NRA personnel of Rwandese origin, invaded Rwanda to overthrow the government of Juvénal Habyarimana. Despite much evidence to the contrary, Museveni has denied any connection to the RPF.

48. Oliver Furley, "Britain from Uganda to Museveni: Blind Eye Diplomacy," in Kumar Rupesinghe (ed.), *Conflict Resolution in Uganda* (Athens: Ohio University Press, 1989), p. 291.

49. Tindigarukayo, "Uganda, 1979–85," p. 617.

50. Omara-Otunnu, *Politics and the Military in Uganda, 1890–1985*, p. 164. Also see *Africa Research Bulletin (Political Series)* (15 August 1985), pp. 7719–7724; and *Africa Confidential* (31 August 1985), p. 4.

51. *Africa Research Bulletin (Political Series)* (15 August 1985), p. 7720; and (15 September 1985), pp. 7760–7761.

52. Colin Legum (ed.), *Africa Contemporary Record: Annual Survey and Documents, 1985–1986* (New York and London: Africana Publishing Company, 1987), pp. B-471–B-472.

53. Quoted in Yoweri K. Museveni, *What Is Africa's Problem?* (Kampala: NRM Publications, 1992), pp. 17–18.

54. The DP Mobiliser's Group rally, which was organized by its leader Michael Kaggwa, was scheduled for 8 May 1993. Hundreds of police prevented the rally from taking place by sealing off central Kampala, setting up roadblocks, and flying helicopters over the rally site. On 14 November 1993, police used similar tactics to stop another DP Mobiliser's Group rally in Kampala. For additional information on the rally, see *Uganda Confidential* (15–31 May 1993), p. 5; *Exposure* (May 1993), pp. 10–12; and *Kenya Times* (15 November 1993), p. 8. For assessments of the Ugandan government's relationship with the press, see *New Vision* (25 January 1992), p. 9. It is interesting to note that in August 1993, the Ugandan cabinet sought to muzzle the press by ordering government departments to stop placing paid notices in newspapers deemed "too critical." For more information, see *Daily Nation* (3 September 1993), p. 10.

55. *Uganda 1986–1991: An Illustrated Review* (Kampala: Fountain Publishers, [1991]), p. 15.

56. Cherry Gertzel, "Uganda's Continuing Search for Peace," *Current History,* Vol. 89, No. 547 (May 1990), pp. 231–232.

57. The text of this bill, known as Legal Notice No. 1 of 1986 (Amendment No. 2) Bill 1989, is contained in "NRC Extends NRM Interim Period," *Resistance* (February 1990), pp. 3–5. Also, for a background summary of this issue, see *New Vision* (5 October 1989), pp. 6–7. Also see *Africa Confidential* (22 January 1993), pp. 1–2.

58. For an example of the opposition's attitude toward the proposed 1994–1995 general election, see Uganda National Democratic Organization (UNDO), "How Democratic Will Museveni's Proclaimed 1994/95 General Elections Be?" (unpublished paper in my possession). Also see *Africa Confidential* (17 April 1992), pp. 3–4; *New African* (August 1989), p. 21. For Museveni's views on multiparty democracy, see *New Vision* (28 June 1991), pp. 8–9; *Africa Analysis* (6 March 1992), p. 2; and *New Vision* (24 March 1992), pp. 8–9. Also see *New Vision* (12 March 1992), p. 4, for an editorial defending the RC elections that took place on 29 February 1992. "Election 92," as it was known in Uganda, allowed Ugandans to vote for members of RCs 1–5. Also see *New Vision* (20 March 1992), pp. 1, 16.

59. *Africa Analysis* (20 August 1993), p. 6. The elections were originally scheduled for December 1993. However, the Ugandan government postponed them ostensibly because it lacked the funds to conduct nationwide elections.

60. Quoted in *New Vision* (13 July 1994), p. 14.

61. For information regarding continued rebel activities, see *New Vision* (1 February 1992), p. 16; and (11 April 1992), p. 1. Reports of NRA counterinsurgency activities are contained in Foreign Broadcast Information Service, *Daily Report: Sub-Saharan Africa* (14 January 1992), p. 5; and (21 January 1992), p. 12.

62. *Daily Telegraph* (19 August 1993), p. 9; *Daily Topic* (13 January 1994), p. 1. Also see *Daily News* (3 December 1993), p. 6. It is interesting to note that former president Tito Okello, who returned to Uganda on 21 November 1993 after eight years in exile, conducted negotiations for the government.

63. The Association of the Bar of the City of New York, *Uganda at the Crossroads: A Report on Current Human Rights Conditions* (New York: Association of the Bar of the City of New York, 1991), pp. 16–25.

64. Amnesty International, *Uganda: Human Rights Violations by the National Resistance Army* (New York: Amnesty International, 1991).

65. Amnesty International, *Uganda: The Failure to Safeguard Human Rights* (London: Amnesty International, 1992), pp. 82–86. Also see Amnesty International, *The 1993 Report on Human Rights Around the World* (Alameda, Calif.: Hunter House, 1993), p. 296.

Chapter 4

1. *New Vision* (23 July 1991), p. 8. It is interesting to note that in the weeks before the census was conducted, the Tanzanian newspaper *Daily News* quoted Ugandan government officials as having expressed fears that the census "could be hindered by ongoing civil strife in some parts of the country." *Daily News* (29 December 1990), p. 2. During the census, Uganda's semiofficial newspaper—*New Vision* (15 January 1991), pp. 1, 12—claimed that antigovernment rebels had disrupted the census in at least two-thirds of Soroti District. Such reports obviously cast doubt on the government's contention that the census had been a "fairly accurate count." By admitting that a reliable census could not be carried out, the government would have had to acknowledge that the rebel problem in north and northeastern Uganda was far worse than revealed in official assessments.

2. *New Vision* (23 July 1991), p. 8.

3. *New Vision* (15 August 1990), p. 12. The Republic of Uganda, *Background to the Budget, 1991–1992* (Kampala: Ministry of Planning and Economic Development, 1991), p. 44.

4. *New Vision* (4 May 1992), pp. 8–9. Also see *New Vision* (18 July 1992), p. 4.

5. For an analysis of the role of these languages in Uganda, see Ruth G. Mukama, "Recent Developments in the Language Situation and Prospects for the Future," in Holger Bernt Hansen and Michael Twaddle (eds.), *Changing Uganda: The Dilemmas of Structural Adjustment and Revolutionary Change* (Athens: Ohio University Press, 1991), pp. 334–350.

6. For a discussion of various aspects of the linguistic issue, see *New Vision* (8 May 1990), pp. 1, 8; (28 November 1990), pp. 8–9; (9 May 1990), p. 4; and (12 March 1991), p. 5.

7. For the most recent and most cogent examination of all aspects of the ethnicity issue, see Kenneth Ingham, *Politics in Modern Africa: The Uneven Tribal Dimension* (London and New York: Routledge, 1990), pp. 11–40. This section is based on Ingham's work.

8. For a full account of the coronation, see Alex Shoumatoff, "The Coronation of King Ronnie," *New York Times Magazine* (17 October 1993), pp. 24–25, 35–36.

9. Ingham, *Politics in Modern Africa*, p. 38.

10. Rita M. Byrnes, "Uganda: The Society and Its Environment," in Rita M. Byrnes (ed.), *Uganda: A Country Study* (Washington, D.C.: USGPO, 1992), p. 70.

11. Buganda at that time was divided into ten chieftainships (*sazas*). Under terms of the agreement, which was negotiated on 20 April 1892 by Captain (later Sir) Frederick Lugard, the Protestants were assigned six *sazas*, the Muslims three, and the Catholics one.

12. Holger Bernt Hansen, *Mission, Church and State in a Colonial Setting: Uganda, 1890–1925* (New York: St. Martin's Press, 1984), p. 439.

13. *Uganda, 1986–1991: An Illustrated Review* (Kampala: Fountain Publishers, [1991]), pp. 6–7.

14. Foreign Broadcast Information Service, *Daily Report: Sub-Saharan Africa* (25 March 1991), p. 8; (15 November 1991), p. 6; *New Vision* (15 November 1991), p. 4; and (2 November 1991), pp. 1, 16.

15. For an account of his activities, see John Milner Gray, "Ahmed bin Ibrahim: The First Arab to Reach Buganda," *Uganda Journal,* Vol. 2, No. 2 (September 1947), pp. 80–97.

16. Alison Butler Herrick et al. (eds.), *Uganda: A Country Study* (Washington, D.C.: USGPO, 1981), p. 115. Also see Hansen, *Mission, Church and State in a Colonial Setting,* pp. 224–225.

17. Uganda Protectorate, *Annual Report of the Education Department, 1951* (Entebbe: Government Printer, 1952), p. 3; O. W. Furley and Tom Watson, *A History of Education in East Africa* (New York: NOK Publishers, 1978), pp. 186, 189, 271; and Cooper F. Odaet, *Implementing Educational Policies in Uganda* (Washington, D.C.: World Bank, 1990), p. 3.

18. Odaet, *Implementing Educational Policies in Uganda,* p. ii. Also see Stephen P. Heyneman, "Education During a Period of Austerity: Uganda, 1971–1981," *Comparative Education Review,* Vol. 27, No. 3 (October 1983), pp. 403–404.

19. The teacher shortage was the greatest problem in the educational field confronting the newly established Museveni regime. In 1988, 56 percent of primary school teachers were untrained; in secondary schools 40 percent of the teachers were untrained or under-trained; and 20 to 25 percent of the faculty in the teacher colleges were untrained. In March 1990, only 437 out of 845 posts were filled at Makerere University. Staff shortages were exacerbated by low salaries (in 1988, a primary school teacher received about $3 a month) and by the fact that teacher salaries often were several months in arrears. For additional information, see W. Senteza Kajubi, "Revolution in Education? Tackling the Diploma Disease," in Hansen and Twaddle, *Changing Uganda,* p. 323.

20. Byrnes, "Uganda: The Society and Its Environment," p. 84. Also see *New Vision* (5 July 1991), pp. 8–9, 14; and (5 February 1991), pp. 6–7.

21. Senteza Kajubi, "Revolution in Education?" pp. 329–330.

22. *New Vision* (9 September 1992), pp. 10–11. Also see *New Vision* (30 September 1992), pp. 10–11.

23. Ibid. (27 November 1992), p. 20.

24. Ibid.

25. Ibid. (1 September 1992), pp. 1, 20. Under the terms of "The Support to Uganda Primary Education Reform Programme," the United States will provide $83 million in balance of payments support and $25 million in project assistance.

26. For a background study on the impact of the AIDS epidemic on Uganda, see George C. Bond and Joan Vincent, "Living on the Edge: Changing Social Structures in the Context of AIDS," in Hansen and Twaddle, *Changing Uganda,* pp. 113–129. A more personalized account is Ed Hooper, Slim: A Reporter's Own Story of AIDS in East Africa (London: Bodley Head, 1990).

27. For an essential account of Uganda's early medical history, see Sir Albert R. Cook, *Uganda Memories (1897–1940)* (Kampala: Uganda Society, 1945). Other useful sources include W. D. Foster, *The Early History of Scientific Medicine in Uganda* (Nairobi: East African Literature Bureau, 1970); and Ann Beck, *A History of the British Medical Administration of East Africa, 1900–1950* (Cambridge: Harvard University Press, 1970).

28. R.M.A. van Zwanenberg with Anne King, *An Economic History of Kenya and Uganda, 1800–1970* (London: Macmillan Press, 1977), pp. 7, 12.

29. The various species of tsetse fly transmit trypanosomiasis (sleeping sickness in humans and nagana in animals). Usually the disease is checked by removing all people and livestock from infected areas, thereby breaking the connection between tsetse flies and humans. For an account of the sleeping sickness campaign, see Beck, *A History of British Medical Administration of East Africa*, pp. 43–47. Bell mistakenly believed that the spread of sleeping sickness could be stopped by breaking one of the links in the transmission of the disease. Since the tsetse fly could not be destroyed, Bell decided to remove people from tsetse fly–infested regions. Scientific evidence later proved Bell wrong. Also see Foster, *Early History of Scientific Medicine in Uganda*, pp. 93–107.

30. For a history of this unit, see G. J. Keane and D. G. Tomblings, *The African Native Medical Corps in the East African Campaign* (London: Richard Clay and Sons, 1926).

31. For further information, see Cook, *Uganda Memories*, p. 353.

32. For a brief account of Amin's impact on Uganda's medical establishment, see Henry Kyemba, *A State of Blood* (New York: Ace Books, 1977), pp. 129–132. From 1974 to 1977, the author served as Amin's minister of health. Also see James Tumusiime (ed.), *Uganda 30 Years 1962–1992* (Kampala: Fountain Publishers, [1992]), p. 112.

33. *Uganda, 1986–1991: An Illustrated Review*, p. 74.

34. Byrnes, "Uganda: The Society and Its Environment," p. 88.

35. Quoted in the *Chicago Tribune* (17 August 1993), p. 4. Also see *Uganda, 1986–1991: An Illustrated Review*, p. 76; and *Boston Globe* (22 July 1993), p. 10.

36. *New Vision* (11 August 1989), pp. 6–7.

37. Ibid.

38. *Africa News* (20 February 1989), p. 5.

39. *New Vision* (17 November 1993), p. 1; (4 April 1992), p. 1; and (19 June 1992), p. 13.

40. *World Bank Watch* (9 March 1992), p. 4.

41. *Independent* (14 January 1994), p. 12.

42. Quoted in Furley and Watson, *A History of Education in East Africa*, p. 281.

43. Grace Akello, *Self Twice-Removed Ugandan Woman* (London: Calverts Press, 1990), p. 8; and Furley and Watson, *A History of Education in East Africa*, p. 282.

44. Quoted in Furley and Watson, *A History of Education in East Africa*, p. 10.

45. Ibid., pp. 10–11.

46. Byrnes, "Uganda: The Society and Its Environment," p. 82.

47. Quoted in Akello, *Self Twice-Removed Ugandan Woman*, p. 11. Also, for a text of Obote's charter, see James H. Mittelman, *Ideology and Politics in Uganda: From Obote to Amin* (Ithaca and London: Cornell University Press, 1975), pp. 271–283.

48. Ironically, Idi Amin was the first Ugandan president to appoint a woman minister. However, Elizabeth Bagaaya Nyabongo lost her portfolio as foreign minister after falling afoul of Amin, who justified discharging Nyabongo by claiming she had violated her office and had insulted the Ugandan people by having sexual relations with a European man in a restroom at Orly Airport, Paris. For additional information, see Elizabeth Nyabongo, *Elizabeth of Toro: The Odyssey of an African Princess* (New York: Simon and Schuster/Touchstone, 1989).

49. *New Vision* (4 March 1992), p. 13.

50. Ibid. (26 September 1990), p. 8. It is interesting to note that the 1967 Constitution, which remained in effect in early 1993, does not outlaw discrimination on grounds of sex.

51. Ibid.

52. *New York Times* (24 February 1991), p. 10; and *New Vision* (17 December 1991), p. 8.

53. Quoted in *New Vision* (8 January 1992), p. 9.

54. Sir Apolo Kagwa, *The Kings of Buganda* (Nairobi: East African Publishing House, 1971).

55. *New Vision* (17 January 1991), pp. 8–9; and (5 February 1992), pp. 8–9. Also see the Republic of Uganda, *Background to the Budget, 1990–1991* (Kampala: Ministry of Planning and Economic Development, 1990), p. 149.

56. *Star* (25 November 1989), pp. 4–5. Also see the *Washington Post* (9 November 1989), pp. E-1–E-2; and *New Vision* (10 March 1990), p. 3.

57. *New Vision* (3 December 1990), p. 16; (15 December 1990), p. 4; and (9 January 1991), pp. 6–7.

58. *New Vision* (31 March 1991), pp. 8–9; and *Africa Analysis* (23 July 1993), p. 23.

Chapter 5

1. Kenneth Ingham, *Obote: A Political Biography* (London and New York: Routledge, 1994), p. 79; and Vijay Gupta, *Obote: Second Liberation* (New Delhi: Vikas Publishing House, 1983), p. 133.

2. World Bank, *Uganda: Growing Out of Poverty* (Washington, D.C.: World Bank, 1993), p. ix; *Africa Research Bulletin (Economic Series)* (16 November–15 December 1993), pp. 11483–11484. On 6 June 1967, Uganda, Kenya, and Tanzania signed a treaty that resulted in the creation of the EAC as a common services organization to operate enterprises such as railways, harbors, telecommunications and the post, and an airline. To facilitate regional integration, the EAC distributed East Africa's resources equitably among the three member nations. Thus, Kenya became the headquarters of East African Airways and the East Africa Railways and Tanzania managed the headquarters of the EAC and the Harbours Commission. For its part, Uganda operated the headquarters for the East African Posts and Telecommunications and the East African Development Bank. The EAC also devised a tax system to protect certain industries in Uganda and Tanzania against similar ones in Kenya, which had a stronger industrial and manufacturing infrastructure. Despite these efforts, Kenya, which during the early 1970s became the financial, economic, and political hub of East Africa, easily dominated the EAC's activities.

3. Quoted in *Uganda, 1986–1991: An Illustrated Review* (Kampala: Fountain Publishers, [1991]), p. 1; and *New Vision* (27 August 1992), p. 8.

4. Ministry of Planning and Economic Development, *Background to the Budget, 1988–1989* (Kampala: Ministry of Planning and Economic Development, 1988), pp. 6–7. Also see *New Vision* (27 August 1992), pp. 8–9.

5. E. O. Ochieng, "Economic Adjustment Programmes in Uganda 1985–9," in Holger Bernt Hansen and Michael Twaddle (eds.), *Changing Uganda: The Dilemmas of Structural Adjustment and Revolutionary Change* (Athens: Ohio University Press, 1991), pp. 53, 56.

6. For background material on the IMF's activities in Uganda, see John Loxey, "The IMF, the World Bank and Reconstruction in Uganda," in Bonnie K. Campbell (ed.), *Structural Adjustment in Africa* (New York: St. Martin's Press, 1989), pp. 56–57, 67–91.

7. Ministry of Planning and Economic Development, *Background to the Budget, 1991–1992* (Kampala: Ministry of Planning and Economic Development, 1991), pp. 17–18. Also see *African Business* (August 1990), p. 38; *New Vision* (1 August 1990), pp. 6–7; and *African Economic Digest* (4 May 1992), p. 10.

8. Uganda has concluded barter trade protocols with numerous nations, including Cuba, Libya, Algeria, Egypt, Spain, the former Soviet Union, and India. Kampala has also concluded similar agreements with many multinational corporations, including BAT International, Lonrho, and Afrex Commodities. For additional information on barter trade, see *New Vision* (21 November 1989), pp. 6–7.

9. The ICO was a consortium of coffee-producing countries that established international production quotas and prices.

10. Economist Intelligence Unit, *Country Report: Uganda, Ethiopia, Somalia, Djibouti,* No. 2 (London: Economist Intelligence Unit, 1992), p. 19.

11. *African Business* (November 1992), p. 18.

12. Ministry of Planning and Economic Development, *Background to the Budget 1991–1992,* p. 48; and *New Vision* (1 July 1992), p. 4.

13. *New Vision* (22 August 1989), pp. 6–7. For an assessment of the impact on the collapse of the ICO agreement on Uganda, see *New Vision* (26 January 1990), pp. 9–10. Other relevant items include *Africa Analysis* (10 November 1989), pp. 1–2; *African Economic Digest* (11 December 1989), p. 4; and *New African* (November 1991), p. 33.

14. *New Vision* (11 April 1990), p. 1; and (12 April 1990), p. 4.

15. *Africa Research Bulletin (Economic Series)* (16 July–15 August 1991), p. 10477.

16. *African Economic Digest* (27 July 1992), p. 5.

17. Linda van Buren, "Economy," in *Africa South of the Sahara 1993* (London: Europa Publications, 1992), p. 901. Also see *Guardian* (23 April 1992), p. 13; and *Africa Research Bulletin (Economic Series)* (16 July–15 August 1991), p. 10477. It should be pointed out, however, that the 4.3-ton cotton harvest in 1990 represented a 34 percent increase over the 3.2-ton harvest in 1989. According to Minister of Finance Crispus Kiyonga, the 1991 cotton crop reflected about a 100 percent increase over the 1990 crop.

18. Economist Intelligence Unit, *Country Profile, 1993–94: Uganda* (London: Economist Intelligence Unit, 1994), p. 16; Ministry of Planning and Economic Development, *Background to the Budget, 1991–1992,* pp. 51–52, 64; and *African Economic Digest* (8 February 1993), p. 12.

19. *Africa Confidential* (27 February 1980), p. 6. Mitchell Cotts, a British company, had owned three groups of tea estates in Uganda until Idi Amin nationalized them.

20. *Africa Research Bulletin (Economic Series)* (16 March–15 April 1991), p. 10347; and Ministry of Planning and Economic Development, *Background to the Budget, 1991–1992,* p. 52.

21. *Africa Research Bulletin (Economic Series)* (16 March–15 April 1991), p. 10347; Ministry of Planning and Economic Development, *Background to the Budget, 1990–1991* (Kampala: Ministry of Planning and Economic Development, 1990), p. 54; and Economist Intelligence Unit, *Country Profile, 1993–94,* p. 17.

22. *Africa Research Bulletin (Economic Series)* (16 July–15 August 1991), p. 10477.

23. Buren, "Economy," p. 901. Also see Ministry of Planning and Economic Development, *Background to the Budget, 1992–1993* (Kampala: Ministry of Finance and Economic Planning, 1992), p. 43.

24. Ministry of Planning and Economic Development, *Background to the Budget, 1991–1992,* p. 58.

25. *New Vision* (13 April 1992), p. 9.

26. For information about the origin of the Nile perch problem, see *Africa* (January 1986), p. 53. For more recent assessments, see the *Washington Post* (7 July 1992), p. A-15; *New Vision* (23 August 1991), pp. 8–9; and (8 September 1992), p. 4.

27. *New Vision* (24 November 1992), p. 4.

28. Ibid. (6 March 1990), p. 4.

29. Buren, "Economy," p. 902; Economist Intelligence Unit, *Country Profile: Uganda, 1988–89* (London: Economist Intelligence Unit, 1988), p. 19; and Ministry of Planning and Economic Development, *Background to the Budget, 1991–1992*, p. 87.

30. Economist Intelligence Unit, *Country Profile: Uganda, 1988–89*, p. 17.

31. Ministry of Planning and Economic Development, *Background to the Budget, 1988–1989*, p. 62.

32. *Africa Analysis* (28 April 1989), p. 8.

33. Economist Intelligence Unit, *Country Profile: Uganda, 1988–89*, p. 18; and Economist Intelligence Unit, *Country Report: Uganda, Ethiopia, Somalia, Djibouti*, No. 3 (London: The Economist Intelligence Unit, 1991), p. 24.

34. Quoted in *Africa Analysis* (13 October 1989), p. 8. Also see *Financial Times* (5 April 1990), pp. 1, 8; *New Vision* (25 June 1990), pp. 1, 12; *New Vision* (14 February 1992), pp. 1, 16; and Nancy Clark, "Uganda: The Economy," in Rita M. Byrnes (ed.), *Uganda: A Country Study* (Washington, D.C.: USGPO, 1992), p. 125.

35. Economist Intelligence Unit, *Country Profile: Uganda, 1988–89*, p. 21.

36. Buren, "Economy," p. 903; and Ministry of Planning and Economic Development, *Background to the Budget, 1991–1992*, p. 110.

37. *New Vision* (29 July 1991), p. 16; and (13 February 1992), pp. 1, 16.

38. Ibid. (29 August 1991), p. 17; (8 January 1992), pp. 1, 16; and *Africa Analysis* (6 March 1992), p. 7.

39. *New Vision* (11 May 1992), p. 9; and *African Economic Digest* (23 March 1992), p. 14.

40. *New Vision* (20 February 1992), p. 4; and (8 June 1992), p. 16. Also see *Africa Analysis* (6 March 1992), p. 7; and *African Economic Digest* (23 March 1992), p. 14.

41. *Uganda, 1986–1991: An Illustrated Review*, p. 55.

42. *Africa Analysis* (21 July 1989), p. 5; and (12 August 1991), p. 14.

43. *New Vision* (13 February 1990), pp. 1, 12.

44. Ibid. (27 August 1991), p. 16; (31 July 1991), pp. 1, 20; and *Africa Events* (June 1991), p. 43.

45. *New Vision* (27 February 1992), pp. 1, 16; and (22 October 1992), p. 2.

46. Clark, "Uganda: The Economy," p. 133; Ministry of Planning and Economic Development, *Background to the Budget, 1991–1992*, p. 224; and *New Vision* (9 August 1994), p. 4.

47. Ministry of Planning and Economic Development, *Background to the Budget, 1991–1992*, pp. 112–113.

48. J. G. Williams, *National Parks of East Africa* (Lexington, Mass.: Stephen Greene Press, 1981), p. 130.

49. *Africa Events* (October 1990), pp. 34–35; and *Financial Times* (31 January 1990), pp. 1, 6.

50. *New Vision* (2 August 1991), p. 2; and (18 July 1991), pp. 1, 16; *Africa Analysis* (20 September 1991), p. 14; and *Los Angeles Times* (15 September 1991), p. A6.

51. In the early 1970s, the elephant population was approximately 15,000. Poaching by soldiers, guerrillas, and private citizens reduced the elephant to the verge of extinction in Uganda. For a more recent assessment of the status of Uganda's fauna, see *New Vision* (27 May 1991), p. 3.

52. In the early 1990s, for example, a trip from Kampala to Queen Elizabeth National Park, a distance of about 240 miles, took nine hours over very poor roads. For a discussion of Uganda's attempts to resolve some of these problems see *African Economic Digest* (3 June 1991), p. 23.

53. Ibid. (20 September 1993), p. 11.

54. *New Vision* (25 June 1991), p. 4.

55. Ibid. (1 July 1992), p. 8.

56. Ministry of Planning and Economic Development, *Background to the Budget, 1991–1992*, p. 22; and *New Vision* (1 July 1992), p. 9.

57. *Africa Research Bulletin (Economic Series)* (16 May–15 June 1992), p. 10885; and *Africa Analysis* (19 July 1992), p. 12.

58. *Africa Analysis* (6 September 1991), p. 13. According to this report, "there are no systematic records to monitor all the aid received by the government." The World Bank claimed that presumably as of mid-1991, Uganda had received a total of $1.978 billion in economic aid; however, the UNDP maintained that the figure should be $2.167 billion. The Ministry of Finance and the Ministry of Planning and Economic Development, both of which coordinate external aid, also "have been unsuccessful" in gathering information about how much foreign assistance Uganda has received. According to the World Bank, statistics kept by the Bank of Uganda are "neither reliable nor up to date."

59. *New Vision* (8 May 1992), pp. 1, 16; and *African Economic Digest* (1 June 1992), p. 12.

60. Although Angola, Burundi, Comoros, Djibouti, Seychelles, Somalia, and Zimbabwe failed to send representatives to the PTA summit in Kampala, they were eligible to sign the treaty.

61. *New African* (January 1994), p. 23.

62. Clark, "Uganda: The Economy," pp. 143–144.

63. *Africa Research Bulletin (Economic Series)* (31 May 1990), p. 9933.

64. *African Business* (June 1990), p. 56.

65. *African Economic Digest* (1 June 1992), p. 17.

66. Quoted in *Africa Research Bulletin (Economic Series)* (16 November–15 December 1993), p. 11484. Also see *Africa Events* (February 1994), pp. 17–27; *African Business* (February 1994), pp. 15–16; and Foreign Broadcast Information Service, *Daily Report: Sub-Saharan Africa* (2 December 1993), p. 3.

Chapter 6

1. For a recent survey of Anglo-Ugandan relations, see Oliver Furley, "Britain and Uganda from Amin to Museveni: Blind Eye Diplomacy," in Kumar Rupesinghe (ed.), *Conflict Resolution in Uganda* (Athens: Ohio University Press, 1989), pp. 275–294.

2. Ibid., p. 282.

3. Ibid., p. 284. Also see Colin Legum (ed.), *Africa Contemporary Record: Annual Survey and Documents, 1981–1982* (New York: Africana Publishing Company, 1982), pp. B-309, B-315.

4. Colin Legum (ed.), *Africa Contemporary Record: Annual Survey and Documents, 1986–1987* (New York and London: Africana Publishing Company, 1987), p. B-473.

5. *New Vision* (10 March 1990), p. 1.

6. Gad W. Toko, *Intervention in Uganda: The Power Struggle and Soviet Involvement* (Pittsburgh: University of Pittsburgh, 1979), pp. 25–30. Also see Thomas P. Ofcansky,

"Uganda: National Security," in Rita M. Byrnes (ed.), *Uganda: A Country Study* (Washington, D.C.: USGPO, 1992), pp. 216–217.

7. "Uganda," in Colin Legum (ed.), *Africa Contemporary Record: Annual Survey and Documents, 1972–1973* (New York: Africana Publishing Company, 1973), p. B-284.

8. Three hostages and the Israeli commander died during the raid. Mrs. Dora Bloch, an elderly hostage who earlier had been taken to Mengo Hospital for medical attention, was never seen again. Most observers believe that the Ugandan authorities murdered her after the raid. For further information on the raid, see "Uganda," in Colin Legum (ed.), *Africa Contemporary Record: Annual Survey and Documents, 1976–77* (New York: Africana Publishing Company, 1977), p. B-386. For the Israeli perspective on the raid, see Yehuda Ofer, *Operation Thunder: The Entebbe Raid* (Harmondsworth: Penguin, 1976); or Uri Dan, *Operation Uganda* (Jerusalem: Keter Publishing House, 1976).

9. On 23 November 1964, five United States Air Force C-130s, escorted by B-26s, air-dropped a Belgian Paracommando Regiment on the Stanleyville Airport. Within hours, the Belgians freed nearly 2,000 Europeans who had been held hostage by Congolese guerrillas known as Simbas. Most independent African nations condemned the raid as yet another example of the West's neocolonial attitude toward Africa. For additional information on the raid, see Major Thomas P. Odom, *Dragon Operations: Hostage Rescues in the Congo, 1964–1965* (Fort Leavenworth, Kans.: U.S Army Command and General Staff College, 1988).

10. James H. Mittelman, *Ideology and Politics in Uganda: From Obote to Amin* (Ithaca and London: Cornell University Press, 1975), p. 104.

11. Officially, the United States closed the embassy because of Amin's opposition to the Vietnam War and the danger to U.S. personnel in Uganda. The U.S. ambassador to Uganda, Thomas Melady, urged the State Department to justify the embassy's closing by saying that Amin "was a murderer and he had engaged in torture and we would not dignify that kind of government with the presence of a U.S. diplomatic mission." Quoted in United States Senate Committee on Foreign Relations, *Uganda: The Human Rights Situation* (Washington, D.C.: USGPO, 1978), p. 99. For background on the boycott, see *Africa News* (14 August 1978), pp. 8–9.

12. Colin Legum (ed.), *Africa Contemporary Record: Annual Survey and Documents, 1982–1983* (New York: Africana Publishing Company, 1984), p. B-323; and Colin Legum (ed.), *Africa Contemporary Record: Annual Survey and Documents, 1984–1985* (New York: Africana Publishing Company, 1986), pp. B-404, B-410.

13. Uganda's relationship with Libya has been particularly troublesome for the United States. In January 1989, relations between Kampala and Washington suffered a setback after Uganda condemned the shooting down of two Libyan reconnaissance planes by the U.S. Navy in the Mediterranean. Since then, U.S. officials have repeatedly warned President Yoweri Museveni about the dangers of maintaining ties to Tripoli.

14. Department of State and Defense Security Assistance Agency, *Congressional Presentation for Security Assistant Programs, Fiscal Year 1991* (Washington, D.C.: Department of State and Defense Security Assistance Agency, 1990), p. 279.

15. *Washington Post* (10 September 1992), p. A-8; and *New York Times* (10 September 1992), p. A-10. U.S. authorities released the Ugandan ambassador after he claimed diplomatic immunity.

16. Bruce D. Larkin, *China and Africa, 1949–1970* (Berkeley: University of California Press, 1973), p. 94.

17. Quoted in Colin Legum (ed.), *Africa Contemporary Record: Annual Survey and Documents, 1978–1979* (New York: Africana Publishing Company, 1980), p. B-440.

18. Legum, *Africa Contemporary Record: Annual Survey and Documents, 1981–1982*, p. B-315.

19. *New Vision* (14 February 1992), p. 15.

20. Colin Legum and Marian E. Doro (eds.), *Africa Contemporary Record: Annual Survey and Documents, 1988–1989* (New York and London: Africana Publishing Company, 1992), p. B-433.

21. The Tanzanians captured about 400 Libyans. On 28 November 1979, they were released, thanks largely to Algerian mediation.

22. *Weekly Review* (12 September 1986), p. 14.

23. *New Vision* (4 June 1992), p. 1.

24. See, for example, ibid. (22 February 1992), pp. 1, 16. Also see Legum and Doro, *Africa Contemporary Record, 1988–1989*, p. B-431.

25. *Daily Nation* (17 December 1987), pp. 1, 6, 7.

26. *Weekly Topic* (22 June 1990), p. 9; *New Vision* (30 May 1990), p. 1; and (30 June 1993), p. 2.

27. Catharine Watson, *Exile from Rwanda: Background to an Invasion* (Washington, D.C.: U.S. Committee for Refugees, 1991), p. 11.

28. Ofcansky, "Uganda: National Security," pp. 184, 219.

29. *New Vision* (23 September 1989), p. 4; (25 November 1989), p. 12; (9 July 1991), p. 16; and (10 July 1991), p. 4. *Sunday News* (16 June 1991), p. 1; *Africa Research Bulletin (Economic Series)* (16 February–15 March 1991), p. 10306; *Daily News* (6 August 1991), p. 2; and *Summary of World Broadcasts* (20 February 1990), p. A-2/4.

30. Foreign Broadcast Information Service, *Daily Report: Sub-Saharan Africa* (14 May 1986), p. R-1; and Colin Legum (ed.), *Africa Contemporary Record: Annual Survey and Documents, 1986–1987* (New York and London: Africana Publishing Company, 1988), p. B-480.

31. The MNC was composed of followers of former Zairian prime minister Patrice Lumumba, who was assassinated in 1961. The MNC, which maintained an unknown number of training camps in Uganda, wanted to overthrow Zairian president Mobutu Sese Seko.

32. *Africa Events* (November 1987), p. 14.

33. *New Vision* (12 August 1991), pp. 1, 4, 20; (4 October 1991), pp. 1, 16; (4 November 1991), p. 20; (9 December 1991), p. 20; and (21 March 1992), pp. 1, 16.

34. *New Vision* (23 March 1992), p. 1; and (1 April 1992), p. 16.

35. Additional information about Ugandan-Sudanese relations is contained in Peter Woodward, "Uganda and Southern Sudan 1986–9," in Holger Bernt Hansen and Michael Twaddle (eds.), *Changing Uganda: The Dilemmas of Structural Adjustment and Revolutionary Change* (Athens: Ohio University Press, 1991), pp. 178–186. Also see *African Concord* (23 March 1992), p. 15.

36. Quoted in the *Sudan Times* (14 June 1987), p. 1. The two leaders also condemned apartheid in South Africa, agreed to promote trade exchanges between Uganda and the

Sudan, and promised to increase cultural, technical, and scientific cooperation. Also see *Africa Confidential* (31 March 1989), p. 8.

37. Colin Legum and Marian E. Doro (eds.), *Africa Contemporary Record: Annual Survey and Documents, 1987–1988* (New York and London: Africana Publishing Company, 1989), p. B-454.

38. Many Western observers learned that Museveni secretly provided aid to the SPLA, if for no other reason than to repay the Sudanese government for allowing the UNLA to operate from bases in Equatoria.

39. *Africa Confidential* (15 July 1988), p. 8.

40. Foreign Broadcast Information Service, *Daily Report: Sub-Saharan Africa* (10 April 1989), pp. 20–21.

41. *New Vision* (21 September 1989), pp. 1, 12.

42. For a Sudanese interpretation of the commission's activities, see "Sudan, Uganda Mend Fences," *Sudanow* (October 1989), pp. 34–35.

43. *Defense and Foreign Affairs Weekly* (27 November–3 December 1989), p. 3; *New Vision* (16 November 1989), pp. 1, 12; *Guide* (6 December 1989), p. 4; and *New Vision* (2 January 1990), pp. 1, 12.

44. Foreign Broadcast Information Service, *Daily Report: Sub-Saharan Africa* (23 January 1990), p. 4; *Africa Research Bulletin (Political Series)* (15 May 1990), p. 9646; and *Guide* (3 April 1990), p. 1.

45. The Sudanese military team did not arrive in Uganda until 3 June 1990.

46. *New Vision* (25 November 1991), pp. 1, 16; and (4 December 1991), pp. 1, 16; and Foreign Broadcast Information Service, *Country Report: Sub-Saharan Africa* (26 September 1991), p. 14. On 30 August 1991, three SPLA officers—Lam Akol, Riek Machar, and Gordon Kong Chuol—unsuccessfully attempted to overthrow SPLA leader John Garang. These three formed the so-called Nasir faction. They also accused Garang of being an auto-cratic dictator and unleashing a reign of terror against his real and imagined opponents in southern Sudan. They also supported southern independence whereas Garang favored a greater southern role in a unitary Sudanese state. Later, William Nyoung, Garang's chief of staff, deserted with some of his followers and formed his own faction. By 1993, these two anti-Garang factions had united into the SPLA/United (SPLA/U), which received support from the Sudanese government. By late 1993, clashes between the SPLA/U and Garang's forces in the Kongor region of southern Sudan had proved inconclusive.

47. *New Vision* (11 February 1992), pp. 8–9. Also see U.S. Committee for Refugees, *World Refugee Survey, 1992* (Washington, D.C.: U.S. Committee for Refugees, 1992), p. 54.

48. Interested readers should also consult David Throup, "Kenya's Relations with Museveni's Uganda," in Hansen and Twaddle, *Changing Uganda*, pp. 187–196.

49. See, for example, Foreign Broadcast Information Service, *Daily Report: Sub-Saharan Africa* (6 October 1986), p. R-1. Also see *Observer* (27 April 1986), p. 19.

50. *Weekly Review* (3 October 1986), p. 24; and *New African* (December 1986), pp. 17–18.

51. *Weekly Review* (10 April 1987), pp. 28–29; (8 January 1988), pp. 34–35; and (15 May 1987), p. 25; *Africa Report* (July–August 1987), p. 9; *Indian Ocean Newsletter* (19 December 1987), p. 1; *Guardian* (15 December 1987), p. 8; and *Africa Research Bulletin (Political Series)* (15 January 1988), pp. 8720–8721.

52. *Weekly Review* (22 July 1988), pp. 42–43; *Kenya Times* (20 July 1988), p. 12.

53. Most Western observers agreed that the aircraft had originated in the Sudan.

54. See, for example, *Kenya Times* (22 October 1990), pp. 1, 4, 6, 13; *Sunday Times* (14 October 1990), p. 1; and (25 November 1990), p. 1.

55. *Weekly Topic* (25 January 1991), pp. 1, 14; *Weekly Review* (15 January 1991), pp. 4–8; *Africa Events* (March 1991), p. 7; *Kenya Times* (11 February 1991), pp. 1–2; *Standard* (23 April 1991), pp. 1, 13; and *New Vision* (11 November 1991), p. 1.

56. The situation in western Kenya, which eventually claimed hundreds of lives, involved clashes between government-armed Kalenjins and Luo, Kisi, and Kikuyu farmers. President Daniel arap Moi repeatedly denied any connection with the so-called Kalenjin warriors. For additional information, see *New African* (July 1992), p. 17. Also see *New Vision* (24 March 1992), p. 4.

57. Watson, *Exile from Rwanda: Background to an Invasion,* p. 14. Also see *Africa Confidential* (12 October 1990), pp. 1–2; and *Africa Research Bulletin (Political Series)* (1–31 October 1990), pp. 9874–9878.

58. Most observers agree that there was probably an element of truth in each of these accusations and counteraccusations. It is interesting to note that the RPF's first leader, Fred Rwigyema, had been a major-general in the NRA and had served as chief of staff and vice-minister of defence. His successor, Major Paul Kigame, had been chief of military intelligence in the NRA.

59. Richard Tebere, "Storm and Stress," *Africa Events* (December 1990), p. 9. Also see *New Vision* (12 November 1990), pp. 1, 16.

60. Joel Krieger (ed.), *The Oxford Companion to Politics of the World* (New York and Oxford: Oxford University Press, 1993), p. 645.

61. *Africa News* (24 June 1991), pp. 10–11.

Chapter 7

1. For example, see Phares Mutibwa, *Uganda Since Independence: A Story of Unfulfilled Hopes* (Trenton, N.J.: Africa World Press, 1992), pp. 1–10.

2. Such a view is highly unpopular, especially among Africanists. However, some journalists have addressed the issue of Uganda's inability to govern itself. For example, see David Lamb, *The Africans* (New York: Vintage Books, 1984), pp. 92–95. For a more general and more chilling analysis of the growing chaos in the Third World, see Robert D. Kaplan, "The Coming Anarchy," *Atlantic Monthly,* Vol. 273, No. 2 (February 1994), pp. 44–76.

3. Despite Museveni's repeated claims to have ended the insurgency problem, rebels are still active in Uganda. In February 1993, antigovernment forces launched a series of attacks against villages, vehicles, and individuals in Kitgum District. At approximately the same time, insurgents also conducted operations along the Ugandan-Zairian border and in Murchison Falls National Park.

4. *Africa Confidential* (22 January 1993), pp. 1–2.

5. For additional information on the UNDP program, see *New Vision* (29 December 1992), p. 3.

6. Yoweri Museveni, *Selected Articles on the Uganda War* (Kampala: NRM Publications, 1986), p. 87.

Bibliography

The following bibliography provides additional information about Uganda. Along with the sources cited in individual chapter notes, some of which also are listed here, these materials reflect only a sampling of the available literature about Uganda.

General Works

Apter, David E. *The Political Kingdom in Uganda*. Princeton: Princeton University Press, 1961.

Burke, Fred G. *Local Government and Politics in Uganda*. Syracuse: Syracuse University Press, 1964.

Coupland, Reginald. *East Africa and Its Invaders*. Oxford: Clarendon Press, 1938.

————. *The Exploitation of East Africa 1856–1890*. London: Faber and Faber, 1968.

Hardy, Ronald. *The Iron Snake*. New York: Putnam, 1965.

Hill, M. F. *Permanent Way*. 2 vols. Nairobi: East African Railways and Harbours Administration, 1949, 1957.

Ingham, Kenneth. *A History of East Africa*. London: Longmans, 1962.

————. *The Making of Modern Uganda*. London: George Allen and Unwin, 1958.

Ingrams, Harold. *Uganda: A Crisis of Nationhood*. London: HMSO, 1960.

Jorgensen, Jan Jelmert. *Uganda: A Modern History*. New York: St. Martin's Press, 1981.

Kaberuka, Will. *The Political Economy of Uganda 1890–1979*. New York: Vantage Press, 1990.

Mamdani, Mahmood. *Politics and Class Formation in Uganda*. New York and London: Monthly Review Press, 1976.

Marsh, Zoe. *An Introduction to the History of East Africa*. Cambridge: Cambridge University Press, 1965.

Maxon, Robert M. *East Africa: An Introductory History*. Morgantown: West Virginia University, 1986.

Meister, Albert. *East Africa: The Past in Chains, the Future in Pawn*. New York: Walker and Company, 1966.

Miller, Charles. *The Lunatic Express*. New York: Macmillan Company, 1971.

Ogot, Bethwell A., and J. A. Kiernan (eds.). *Zamani: A Survey of East African History*. Nairobi: East African Publishing House, 1968.

Omara-Otunnu, Amii. *Politics and the Military in Uganda 1890–1985*. New York: St. Martin's Press, 1987.

Sathyamurthy, T. V. *The Political Development of Uganda Nineteen Hundred to Nineteen Eighty-Six*. London: Gower Publishing Company, 1986.

Thomas, Harold B., and Robert Scott. *Uganda*. London: Oxford University Press, 1935.

Trimingham, J. Spencer. *Islam in East Africa*. Oxford: Clarendon Press, 1964.
Uganda, 1986–1991: An Illustrated Review. Kampala: Fountain Publishers, [1991].
Wallis, H. R. *The Handbook of Uganda*. London: Crown Agents for the Colonies, 1913.
Ward, W.E.F., and L. W. White. *East Africa: A Century of Change, 1870–1970*. New York: Africana Publishing Corporation, 1971.
Were, Gideon S., and Derek A. Wilson. *East Africa Through a Thousand Years*. New York: Africana Publishing Corporation, 1968.

Physical Setting

Barnes, J. W. *The Mineral Resources of Uganda*. Entebbe: Government Printer, 1961.
Brown, Leslie. *East African Mountains and Lakes*. Nairobi: East African Publishing House, 1971.
Johnston, Sir Harry Hamilton. *The Uganda Protectorate*. 2 vols. London: Hutchinson, 1902.
Langdale-Brown, I., H. A. Osmaston, and J. G. Wilson. *The Vegetation of Uganda and Its Bearing on Land Use*. Entebbe: Government of Uganda, 1964.
McEwen, A. C. *International Boundaries of East Africa*. Oxford: Clarendon Press, 1971.
Morgan, W.T.W. *East Africa: Its People and Resources*. Nairobi: Oxford University Press, 1971.
Republic of Uganda. Department of Land and Surveys. *Atlas of Uganda*. 2d ed. Kampala: Government Printer, 1967.
Thomas, Harold B., and A. E. Spencer. *A History of Uganda Land and Surveys and of the Uganda Land and Survey Department*. Entebbe: Government Printer, 1938.
Uganda Protectorate. *Mineral Resources of Uganda*. Entebbe: Government Printer, 1942.

Preindependence History

Archer, Geoffrey. *Personal and Historical Memoirs of an East African Administrator*. Edinburgh and London: Oliver and Boyd, 1963.
Ashe, Robert P. *Chronicles of Uganda*. London: Hodder and Stoughton, 1894.
———. *Two Kings of Uganda*. London: Sampson Low, 1890.
Austin, Herbert H. *With MacDonald in Uganda*. London: Edward Arnold, 1903.
Beattie, John. *The Nyoro State*. Oxford: Clarendon Press, 1971.
Bell, Hesketh. *Glimpses of a Governor's Life*. London: Sampson Low, Marston and Company, 1946.
Colvile, Henry E. *The Land of the Nile Springs*. London: Edward Arnold, 1895.
Cook, Albert R. *Uganda Memories, 1897–1940*. Kampala: Uganda Society, 1945.
Cunningham, James F. *Uganda and Its People*. London: Hutchinson, 1905.
Driberg, Jack H. *The Lango: A Nilotic Tribe of Uganda*. London: T. Fisher Unwin, 1923.
Dunbar, A. R. *A History of Bunyoro-Kitara*. Nairobi: Oxford University Press, 1970.
Edel, M. M. *The Chiga of Western Uganda*. New York: Oxford University Press for the International African Institute, 1957.
Galbraith, John S. *Mackinnon and East Africa, 1878–1895: A Study in the "New Imperialism."* Cambridge: Cambridge University Press, 1972.
Ingham, Kenneth. *The Kingdom of Toro in Uganda*. London: Methuen and Company, 1958.

Karugire, Samwiri Rubaraza. *A History of the Kingdom of Nkore in Western Uganda to 1896.* Oxford: Clarendon Press, 1971.

———. *A Political History of Uganda.* London: Heinemann, 1980.

Kiwanuka, M.S.M. Semakula. *A History of Buganda from the Foundation of the Kingdom to 1900.* London: Longman, 1971.

Lamphear, J. *The Traditional History of the Jie of Uganda.* Oxford: Clarendon Press, 1976.

Lardner, E. G. Dion. *Soldiering and Sport in Uganda 1909–1910.* London: Walter Scott Publishing Company, 1912.

Low, Donald A. *Buganda in Modern History.* Berkeley: University of California Press, 1971.

———. *The Mind of Buganda.* Berkeley and Los Angeles: University of California Press, 1971.

———. *Political Parties in Uganda, 1949–62.* London: Athlone Press, 1962.

Low, Donald A., and R. Cranford Pratt. *Buganda and British Overrule.* Nairobi: Oxford University Press, 1970.

Lugard, Frederick. *The Rise of Our East African Empire.* 2 vols. Edinburgh and London: William Blackwood and Sons, 1893.

MacDonald, J.R.L. *Soldiering and Surveying in British East Africa, 1891–1894.* London: Edward Arnold, 1897.

McDermott, P. L. *British East Africa or IBEA.* London: Chapman and Hall, 1895.

Nabudere, D. Wadada. *Imperialism and Revolution in Uganda.* London: Onyx Press, 1980.

O'Brien, T. P. *The Prehistory of Uganda Protectorate.* Cambridge: Cambridge University Press, 1939.

Portal, Gerald. *The British Mission to Uganda in 1893.* London: Edward Arnold, 1894.

Ray, Benjamin C. *Myth, Ritual, and Kingship in Buganda.* Oxford: Oxford University Press, 1991.

Roscoe, John. *The Northern Bantu: An Account of Some Central African Tribes of the Uganda Protectorate.* Cambridge: Cambridge University Press, 1915.

Steinhart, Edward I. *Conflict and Collaboration: The Kingdoms of Western Uganda, 1880–1907.* Princeton: Princeton University Press, 1977.

Thruston, A. B. *African Incidents.* London: John Murray, 1900.

Vandeleur, Seymour. *Campaigning on the Upper Nile.* London: Methuen and Company, 1898.

Wild, John V. *The Story of the Uganda Agreement.* Nairobi: East African Literature Bureau, 1950.

———. *The Uganda Mutiny.* London: Macmillan, 1954.

The Postindependence Period

Avirgan, Tony, and Martha Honey. *War in Uganda: The Legacy of Idi Amin.* Westport, Conn.: Lawrence Hill and Company, 1982.

Bwengye, A.W.F. *The Agony of Uganda: From Idi Amin to Obote.* London: Regency Press, 1985.

Colletta, N. J., and N. Ball. "War to Peace Transition in Uganda," *Finance and Development,* Vol. 30, No. 2 (June 1993), pp. 36–39.

Dodge, C. P., and Magne Raundalen (eds.). *War, Violence, and Children in Uganda.* Oslo: Norwegian University Press, 1987.

Dodge, C. P., and P. D. Wiebe (eds.). *Beyond Crisis: Development Issues in Uganda.* Kampala: Makerere Institute of Social Research, Makerere University and African Studies Association, 1987.

———. *Crisis in Uganda: The Breakdown of Health Services.* Oxford: Pergamon Press, 1985.

Furley, Oliver. *Uganda's Retreat from Turmoil?* London: Centre for Security and Conflict Studies, n.d.

Gingyera-Pinycwa, A. G. *Milton Obote and His Times.* New York: NOK Publishers, 1978.

Hansen, Holger Bernt, and Michael Twaddle (eds.). *Changing Uganda.* Athens: Ohio University Press, 1991.

——— *Uganda Now: Between Decay and Development.* Athens: Ohio University Press, 1988.

Ingham, Kenneth. *Obote: A Political Biography.* London and New York: Routledge, 1994.

Kiwanuka, Semakula. *Amin and the Tragedy of Uganda.* Munich and London: Weltform Verlag, 1979.

Kyobe, V., et al. *Proposed Federal States of Uganda.* 2d ed. Kampala: Makerere University Printery, 1988.

Mamdani, Mahmood. *Imperialism and Fascism in Uganda.* Nairobi: Heinemann Educational Books, 1983.

———. "Uganda in Transition," *Third World Quarterly,* Vol. 10, No. 3 (July 1988), pp. 1155–1181.

Martin, David. *General Amin.* London: Faber and Faber, 1974.

Mazrui, Ali A. *Soldiers and Kinsmen in Uganda.* Beverly Hills, Calif.: Sage Publications, 1975.

Mittelman, James H. *Ideology and Politics in Uganda: From Obote to Amin.* Ithaca and London: Cornell University Press, 1975.

Museveni, Yoweri. *The Path of Liberation.* Kampala: Government Printer, 1989.

———. *Selected Articles on the Uganda Resistance War.* Kampala: NRM Publications, 1985.

———. *Ten Point Programme of NRM.* Kampala: NRM Publications, 1985.

———. *What Is Africa's Problem?* Kampala: NRM Publications, 1992.

Mutibwa, Phares. *Uganda Since Independence: A Story of Unfulfilled Hopes.* Trenton, N.J.: Africa World Press, 1992.

Obote, Milton A. *The Common Man's Charter.* Entebbe: Government Printers, 1970.

Rothchild, Donald, and John W. Harbeson. "Rehabilitation in Uganda," *Current History,* Vol. 80, No. 463 (March 1981), pp. 115–119, 134–138.

Rupesinghe, Kumar. *Conflict Resolution in Uganda.* Athens: Ohio University Press, 1989.

Rupesinghe, Kumar (ed.). *Uganda: Internal Conflict and Its Resolution.* Kampala: The Uganda Bookshop, 1989.

Tindigarukayo, J. "On the Way to Recovery: Museveni's Uganda in 1988," *Canadian Journal of Development Studies,* Vol. 11, No. 2 (1990), pp. 347–357.

Tumusiime, James (ed.). *Uganda: 30 Years, 1962–1992.* Kampala: Fountain Publishers, [1993].

The Economy

Brett, Edwin A. *Colonialism and Underdevelopment in East Africa: The Politics of Economic Change, 1919–1939.* London: Heinemann, 1973.

Commonwealth Team of Experts. *The Rehabilitation of the Economy of Uganda.* 2 vols. London: Commonwealth Secretariat, 1979.

Ehrlich, Cyril. "The Uganda Economy, 1903–1945," pp. 395–475 in Vincent Harlow and E. M. Chilver, eds. *History of East Africa.* Vol. 2. London: Oxford University Press, 1965.

Elkan, Walter. *Crops and Wealth in Uganda.* London: Oxford University Press, 1970.

————. *The Economic Development of Uganda.* London: Oxford University Press, 1961.

Faaland, Just, and Hans-Erik Dahl. *The Economy of Uganda.* Bergen: Christian Michelsen Institute, 1967.

Ford, V.C.R. *The Trade of Lake Victoria.* Nairobi: East African Literature Bureau, 1955.

International Bank for Reconstruction and Development. *The Economic Development of Uganda: Report of a Mission Organized by the International Bank for Reconstruction and Development.* Baltimore: Johns Hopkins Press on Behalf of the International Bank for Reconstruction and Development, 1962.

Jameson, J. D. (ed.). *Agriculture in Uganda.* Oxford: Oxford University Press, 1970.

Lury, D. A. "Dayspring Mishandled: The Uganda Economy, 1945–1960," pp. 212–250 in D. A. Low and Allison Smith, eds. *History of East Africa.* Vol. 3. London: Oxford University Press, 1976.

Mamdani, M. "Uganda: Contradictions of the IMF Programme and Perspective," *Development and Change,* Vol. 21, No. 3 (July 1990), pp. 427–467.

Tothill, John D. *Agriculture in Uganda.* London: Oxford University Press, 1940.

World Bank. *Report on Economic Development in Uganda.* Washington, D.C.: World Bank, 1962.

————. *Uganda: Agricultural Sector Memorandum: The Challenge Beyond Rehabilitation.* Washington, D.C.: World Bank, 1984.

————. *Uganda: Country Economic Memorandum.* Washington, D.C.: World Bank, 1982.

————. *Uganda: Progress Toward Recovery and Prospects for Development.* Washington, D.C.: World Bank, 1985.

Wrigley, Christopher C. *Crops and Wealth in Uganda: A Short Agrarian History.* London: Oxford University Press for Makerere Institute of Social Research, 1973.

Zwanenberg, R.M.A., with Anne King. *An Economic History of Kenya and Uganda, 1800–1970.* London: Macmillan Press, 1977.

Society

Foster, W. D. *The Early History of Scientific Medicine in Uganda.* Nairobi: East African Literature Bureau, 1970.

————. *Sir Albert Cook: A Missionary Doctor in Uganda.* (Newhaven, Sussex, 1978).

Furley, O. W., and Tom Watson. *A History of Education in East Africa.* New York: NOK Publishers, 1978.

Goldthorpe, John E. *An African Elite: Makerere College Students, 1922–1960.* Nairobi: Oxford University Press, 1965.

Heyneman, Stephen P. "Education During a Period of Austerity: Uganda, 1971–1981," *Comparative Education Review*, Vol. 27 (October 1983), pp. 403–413.

Hooper, Ed. "AIDS in Uganda," *African Affairs*, Vol. 86, No. 345 (October 1987), pp. 469–477.

———. *Slim: A Reporter's Own Story of AIDS in East Africa*. London: Bodley Head, 1990.

Kanyeihamba, George W. *Constitutional Law and Government in Uganda*. Nairobi: East African Literature Bureau, 1975.

Ladefoged, Peter, Ruth Glick, and Clive Criper. *Language in Uganda*. London: Oxford University Press, 1972.

Macpherson, Margeret. *They Built for the Future: A Chronicle of Makerere University College, 1922–1962*. Cambridge: Cambridge University Press, 1964.

McGregor, G. P. *King's College, Budo: The First Sixty Years*. London: Oxford University Press, 1967.

Morris, H. F., and James S. Read. *Uganda: The Development of Its Laws and Constitution*. London: Stevens, 1966.

Morris, H. S. *Indians in Uganda*. Chicago: University of Chicago Press, 1968.

Ssekamwa, J. C., and S.M.E. Lugumba. *Educational Development and Administration in Uganda, 1900–1970*. Kampala: Longman Uganda, 1974.

Bibliographies and References

Africa Contemporary Record. New York and London: Africana Publishing Company, annual since 1968–1969.

Byrnes, Rita M. (ed.). *Uganda: A Country Study*. Washington, D.C.: USGPO, 1990.

Collison, Robert L. (ed.). *Uganda*. Oxford and Santa Barbara: Clio Press, 1981.

Gertzel, Cherry. *Uganda: An Annotated Bibliography of Source Materials*. Oxford: Hans Zell, 1991.

Gray, Beverly Ann. *Uganda: Subject Guide to Official Publications*. Washington, D.C.: Library of Congress, 1977.

Herrick, Allison Butler, et al. *Area Handbook for Uganda*. Washington, D.C.: USGPO, 1969.

———. *Uganda: A Country Study*. Washington, D.C.: USGPO, 1981.

Hoben, Susan Y. *A Select Annotated Bibliography of Social Science Materials for Uganda Followed by an Expanded Bibliography*. Washington, D.C.: USAID, 1979.

Hopkins, Terence K., with Perezi Kamunvire. *A Study Guide for Uganda*. Boston: African Studies Center, Boston University, 1969.

Howell, John Bruce. *East African Community: Subject Guide to Official Publications*. Washington, D.C.: USGPO, 1976.

Jamison, Martin. *Idi Amin and Uganda: An Annotated Bibliography*. Westport, Conn.: Greenwood Press, 1992.

Kleinschmidt, Harals (compiler). *Amin Collection, Bibliographical Catalogue of Materials Relevant to the History of Uganda Under the Military Government of Idi Amin Dada*. Heidelburg: P. Kivouvou Verlag-Editions Bantoues, 1983.

Kuria, Lucas, Iris Ragheb, and John Webster (compilers). *A Bibliography on Politics and Government in Uganda*. Syracuse, N.Y.: Syracuse University, Maxwell School of Citizenship and Public Affairs, 1965.

Ofcansky, Thomas P. *British East Africa, 1856–1963: An Annotated Bibliography.* New York and London: Garland Publishing, 1985.

Thurston, Anne. *Guide to Archives and Manuscripts Relating to Kenya and East Africa in the United Kingdom.* 2 vols. Oxford: Hans Zell Publishers, 1990.

Walker, Audrey A. (compiler). *Official Publications of British East Africa, Part 4: Uganda.* Washington, D.C.: Library of Congress, General Reference and Bibliography Division, 1960–1962, 1963.

About the Book and Author

UGANDA, A LANDLOCKED NATION in East Africa, was known during colonial times as the "Pearl of Africa," largely because of its pleasant climate and rich land. For most of the postindependence period, however, Uganda was one of the most brutal and violent nations in Africa. In 1986, a new government seized power, promising to restore internal stability and economic prosperity. Since then, Uganda has gradually become a model for other African states struggling to improve the lives of their citizens.

In this broad survey, Thomas P. Ofcansky examines the political, economic, and social themes that have shaped Ugandan history. He inspects the impact of British colonial rule, investigates the emergence of the independence movement after World War II, and analyzes the factors that contributed to the collapse and decay of Ugandan society after Idi Amin's seizure of power in 1971. The author then explores the successes, failures, and prospects of Uganda's current government. In his conclusion, Ofcansky examines the difficulties facing a nation still divided by ethnic, religious, and regional cleavages and argues that Ugandan leaders must work to establish a society in which all Ugandans benefit or face the prospect of a return to anarchy.

Thomas P. Ofcansky is senior analyst on East Africa for the Department of Defense.

Index

CPSIA information can be obtained at www.ICGtesting.com
Printed in the USA
LVOW092034160812

294501LV00001B/90/A